LECTURES ON
ENGINEERING MECHANICS

STATICS AND DYNAMICS

Black and white print version

STEFAN LINDSTRÖM

LECTURES ON ENGINEERING MECHANICS: STATICS AND DYNAMICS
Stefan Lindström

Black and white print version

ISBN 978-91-981287-3-4
Translated by Eva-Karin Lindström

Other versions of this book, with exercises:
ISBN 978-91-981287-5-8: SI version, Amazon, 2021
ISBN 978-91-981287-8-9: USC version, Amazon, 2021

Contents

I   STATICS   **1**

1   Introduction   **2**
    1.1   Fundamental concepts   2
    1.2   Newton's laws of motion   4
    1.3   Forces in Classical Mechanics   4

2   Force-couple systems   **7**
    2.1   Forces   7
    2.2   Moments and couples   8
    2.3   Force-couple systems   10
    2.4   Planar force-couple systems   12

3   Static equilibrium   **14**
    3.1   Equilibrium equations   14
    3.2   Free-body diagrams   15
    3.3   Multi-body mechanical systems   19

4   Center of mass   **21**
    4.1   Density   21
    4.2   Center of mass   21
    4.3   Center of mass of slender bodies   23
    4.4   Center of gravity   23

5   Distributed and internal forces   **25**
    5.1   Surface and line loads   25
    5.2   Internal forces and couples   27
    5.3   Fluid statics   30

6   Friction   **36**
    6.1   A friction experiment   36
    6.2   Coulomb friction   37
    6.3   Friction in multi-body systems   38
    6.4   Belt friction   39

II PARTICLE DYNAMICS   41

7 Planar kinematics of particles   42
7.1 Rectilinear motion 42
7.2 Curvilinear motion 43
7.3 Kinematic constraints 49

8 Kinetics of particles   51
8.1 Newton's laws of motion 51
8.2 Equations of motion and problem solving 53

9 Work–energy method for particles   57
9.1 Power and Work 57
9.2 Work 57
9.3 Kinetic energy 59
9.4 Conservative forces 60
9.5 Work–energy theorem with potentials 62

10 Momentum and angular momentum of particles   63
10.1 Momentum and impulse 63
10.2 Angular momentum 64
10.3 Systems of particles 66
10.4 Impact 69

11 Harmonic oscillators   73
11.1 Free harmonic oscillators 73
11.2 Forced harmonic oscillators 76

III RIGID BODY DYNAMICS   81

12 Planar kinematics of rigid bodies   82
12.1 Planar motion of rigid bodies 82
12.2 Instantaneous center 86
12.3 Rolling without slipping 88

13 Planar kinetics of rigid bodies   90
13.1 Euler's laws of motion 90
13.2 Euler's first law: the Force equation 91
13.3 Euler's second law: the Moment equation 92
13.4 Flat rigid bodies in planar motion 94

14 Work–energy method for rigid bodies   99
14.1 Power and work from forces and couples 99
14.2 Power sum and work on a rigid body 100

14.3 Kinetic energy 102

14.4 Work–energy theorem 103

15 Impulse relations for rigid bodies   106

15.1 Integral form of Euler's laws 106

15.2 Impulse relations for rigid bodies 107

15.3 Impact with rigid bodies 108

16 Three-dimensional kinematics of rigid bodies   111

16.1 Angular velocity and angular acceleration 111

16.2 Coriolis equation 112

16.3 Velocity and acceleration equations 114

16.4 Systems of rigid bodies 115

17 Three-dimensional kinetics of rigid bodies   118

17.1 The inertia matrix 118

17.2 Angular momentum 121

17.3 Dynamic phenomena 124

APPENDIX   131

A Selected mathematics   132

A.1 Geometry 132

A.2 Vectors 133

A.3 Differentials 137

A.4 Integrals 138

B Quantity, unit and dimension   140

C Tables   144

Index   146

Preface

This textbook gives a concise description of elementary Engineering Mechanics including definitions and theorems. It is suitable for Bachelor's level engineering studies.

It is presumed that the reader is familiar with basic geometry (Appendix A.1), geometric vectors (Appendix A.2), linear systems of equations, ordinary differential equations including differential notation (Appendix A.3), and integrals in several dimensions (Appendix A.4). Besides this, the reader should be acquainted with the concepts of quantity, unit and dimension, and be able to determine whether a physical expression is dimensionally correct (Appendix B).

Acknowledgments

The author wishes to acknowledge his debt to Dr. Peter Schmidt, Dr. Lars Johansson and Dr. Ulf Edlund for their useful feedback.

Dr. Stefan Lindström
Linköping
June 2019

PART I
STATICS

1
Introduction

This chapter introduces some fundamental concepts of Mechanics, and outlines the field of Statics. It is necessary to be familiar with vectors and their properties (Appendix A.2) to grasp the key concepts, and to solve applied problems.

1.1 Fundamental concepts

Bodies and rigid bodies

A *body* occupies a finite region in space, and therefore, it has a volume. A body also has mass, and this mass is assumed to be continuously distributed within the region of the body.

All physical bodies can deform (change their shape). That is, the distance between the material points within the body can change. In some situations this deformation is very small and can be neglected. In such cases, we assume that the shape of the body is unchanging, and we call this a *rigid body model*.

Definition 1.1 (Rigid body). A *rigid body* is a body with the constraint, that the distance between each pair of its material points cannot change.

Particles

A *particle* is a hypothetical object with mass, but without volume. Therefore, all its mass is concentrated to one point. In applications, we sometimes use a *particle model* for a body when its rotation and deformation do not affect the analysis to any great extent. Particularly, we can formulate the following *postulate*[1]:

Postulate 1.2 (Particle). A body, or a part of a body, whose extension is sufficiently small to be neglected in a given situation, can be considered as a particle.[2]

[1] *Postulate* – an unproven statement with experimental support.

[2] J. B. Griffiths. *The theory of classical mechanics*. Cambridge University Press, 1985. ISBN 0-521-23760-2

Position, velocity and acceleration

The position of a point or a particle in space is represented by its *position vector*[3]. We define the position vector of a point $\mathcal{P}$ as $\bar{r} \equiv \overline{\mathcal{OP}}$, where $\mathcal{O}$ denotes the origin of a coordinate system, *e.g.* a rectangular coordinate system with coordinates x, y and z, and corresponding basis vectors $\bar{e}_x$, $\bar{e}_y$ and $\bar{e}_z$. If the position of $\mathcal{P}$ changes with time t, the position vector becomes a vector-valued function (see Appendix A.2)

$$\bar{r}(t) = x(t)\bar{e}_x + y(t)\bar{e}_y + z(t)\bar{e}_z, \tag{1.1}$$

which can be interpreted as a directed path of motion (Fig. 1.1a). The *velocity* of $\mathcal{P}$ is defined as

$$\bar{v}(t) \equiv \frac{\mathrm{d}\bar{r}}{\mathrm{d}t} = \dot{x}\bar{e}_x + \dot{y}\bar{e}_y + \dot{z}\bar{e}_z, \tag{1.2}$$

and it is oriented in the tangent direction of the path of motion. A superposed dot over a scalar function denotes the time derivative of that function. The *acceleration* of $\mathcal{P}$ is defined as

$$\bar{a}(t) \equiv \frac{\mathrm{d}\bar{v}}{\mathrm{d}t} = \frac{\mathrm{d}^2\bar{r}}{\mathrm{d}t^2} = \ddot{x}\bar{e}_x + \ddot{y}\bar{e}_y + \ddot{z}\bar{e}_z. \tag{1.3}$$

Thus, it describes the rate of change of the velocity. Two superposed dots over a scalar function denote the second time derivative of that function.

When a point or particle moves at a constant velocity $\bar{v}$, it is said to describe *uniform motion*, and $\bar{r}(t)$ becomes a rectilinear path (Fig. 1.1b). As a special case, when $\bar{v} = \bar{0}$, $\bar{r}(t)$ identifies a fixed point. In both cases, it follows from Eq. (1.3) that $\bar{a} = \bar{0}$.

Forces

When we place two objects sufficiently close to each other, or if they come into contact, the objects can affect each other's motion. If, for instance, one places a magnet near a steel pin, this pin will accelerate towards the magnet. The ability of two bodies to affect each other's motion is called *interaction*[4].

The concept of *force* is introduced to quantify the magnitude and direction of the interactions of an object with its surroundings. An interaction creates a force that makes the object accelerate. Otherwise, the object would remain still, or move uniformly in rectilinear motion. The force is defined through the laws of particle motion, formulated by Sir Isaac Newton in *Principia* (1687).

[3] Also called *radius vector*.

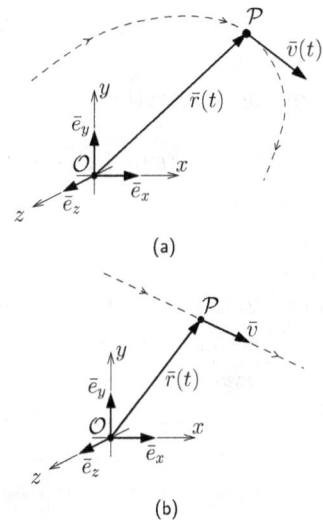

(a)

(b)

Figure 1.1: The path of motion $\bar{r}(t)$ of a point $\mathcal{P}$, with (a) a varying velocity $\bar{v}(t)$, or with (b) a constant velocity $\bar{v}$, and acceleration $\bar{a} = \bar{0}$.

[4] Also called *reciprocal action*.

1.2   Newton's laws of motion

Sir Isaac Newton postulated the following three laws of motion for particles:

1. *Law of inertia*   A particle remains at rest, or moves in a straight line at a constant velocity, as long as the particle does not interact with any other object.

2. *Law of force and acceleration*   For a particle with constant mass m, it holds that

$$\Sigma \bar{F} = m\bar{a}, \tag{1.4}$$

where $\Sigma \bar{F}$ is the vector sum of all forces acting on the particle, and $\bar{a}$ is the acceleration of the particle.

3. *Law of action and reaction*   If a particle exerts a force on another particle, then the latter exerts a force of equal magnitude, but opposite direction, on the former particle.

These laws are treated comprehensively in Chapter 8.

Inertial system

In order to describe motion, a coordinate system first needs to be specified. Newton's laws are only valid for particular coordinate systems called *inertial systems*. If a coordinate system is chosen so that the Law of inertia is valid, then the Law of force and acceleration and the Law of action and reaction become valid in that coordinate system as well. In coordinate systems that rotate or otherwise accelerate relative to an inertial system, Newton's three laws are not valid (Fig. 1.2).

Statics concerns mechanical systems for which all material points describe uniform motion with the same constant velocity in an inertial system.

1.3   Forces in Classical Mechanics

Forces may act on a body if this body is in physical contact with another body. Forces can also act over a distance, *e.g.*, through a gravitational or magnetic field. We measure force in the SI unit of newton (N), or in the USC unit of pound-force (lb$_\mathrm{f}$), where

$$1\,\mathrm{N} = 1\,\frac{\mathrm{kg \cdot m}}{\mathrm{s}^2}, \qquad 1\,\mathrm{lb_f} \approx 4.4482\,\mathrm{N}.$$

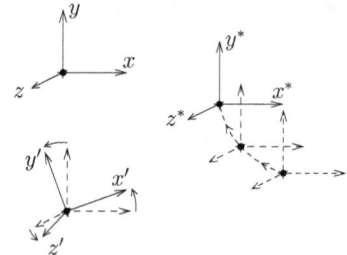

Figure 1.2: Given an inertial system xyz, where the Law of inertia is valid, any coordinate system $x'y'z'$ that rotates relative to this inertial system is not an inertial system. Coordinate systems $x^*y^*z^*$ for which the origin accelerates relative to the inertial system are not inertial systems either.

Universal law of gravitation

According to *Newton's universal law of gravitation*, every pair of particles affect each other with gravitational forces. The gravitational force is a central force of attraction. That is, a pair of particles are drawn towards each other, and the gravitational force acts along a straight line connecting these particles (Fig. 1.3).

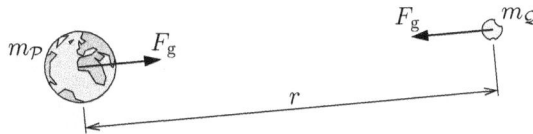

Figure 1.3: Newton's universal law of gravitation for particles applied to the Earth's interaction with the Moon.

Postulate 1.3 (Gravitational force). Between two particles with masses $m_{\mathcal{P}}$ and $m_{\mathcal{Q}}$, respectively, there is an attractive force with magnitude

$$F_{\text{g}} = G_{\text{g}} \frac{m_{\mathcal{P}} m_{\mathcal{Q}}}{r^2}, \tag{1.5}$$

where

$$G_{\text{g}} = 6.67408 \cdot 10^{-11} \, \frac{\text{m}^3}{\text{kg} \cdot \text{s}^2} \approx 3.4397 \cdot 10^{-8} \, \frac{\text{ft}^4}{\text{lb}_{\text{f}} \cdot \text{s}^4} \tag{1.6}$$

is the *gravitational constant* and r is the distance between the particles.

One consequence of the Universal law of gravitation is that a body with mass m near the surface of the Earth is affected by a gravitational force directed towards the center of the Earth. This gravitational force is distributed over the region occupied by the body. However, in many applications gravitation can be modeled as *one* force acting in one point. This force is the *weight* of the body, and has the magnitude mg, where g is the *local gravity constant*[5]. The value of the local gravity constant varies around the world. The *standard gravity*[6]

$$g_{\text{n}} \equiv 9.80665 \, \frac{\text{N}}{\text{kg}} \approx 32.174 \, \frac{\text{ft}}{\text{s}^2}$$

is often used in problem solving.

[5] Also called the *acceleration of gravity*.

[6] Bureau International des Poids et Mesures. The International System of Units (SI), 2006

Contact forces

Two bodies in physical contact with each other interact through *contact forces*. These contact forces are distributed over the touching surfaces of the respective bodies. One example is the force that appears when you press your hand against a wall (Fig. 1.4ab). Your hand exerts a pressure on the wall, which can be represented by a force $\bar{F}$. Conversely, the wall exerts a force $-\bar{F}$ on your hand according to the Law of action and reaction.

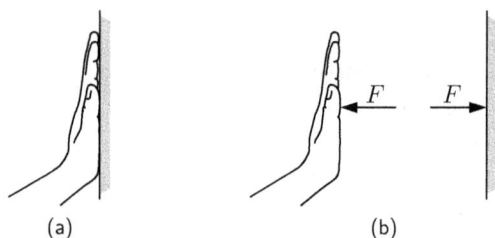

Figure 1.4: (a) Your hand presses against a wall. (b) Your hand and the wall are subjected to equally large but oppositely directed contact forces.

Spring forces

Spring forces appear when a spring is elongated or compressed. When a spring is not affected by any force, it assumes its *natural length* ℓ_0 (Fig. 1.5a). If oppositely directed forces of equal magnitude F_s act in both ends of this spring, it will change its length to ℓ (Fig. 1.5b). For a *linear spring*, the relation

$$F_s = k(\ell - \ell_0) \tag{1.7}$$

holds, where k is the *spring constant* with the SI units of N/m and the USC units of lb_f/ft.

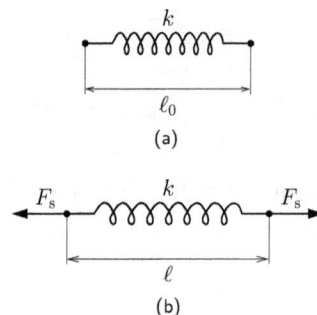

Figure 1.5: (a) Unloaded spring with its natural length. (b) The same spring when elongated by a tensile force.

2

Force-couple systems

2.1 Forces

A body interacts with its surroundings through *forces*. These can be *body forces* that act over the region in space occupied by the body. Gravitational and electromagnetic forces are examples of body forces. A body can also be affected by *contact forces* that are distributed across the surface of the body (Fig. 2.1). For rigid bodies, the body and contact forces can be represented by concentrated forces that act in distinct points on the rigid body.

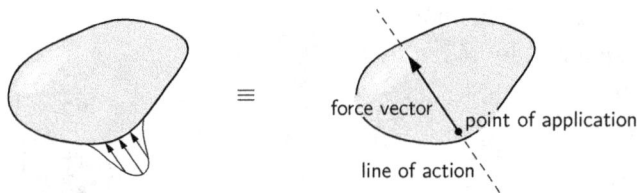

Figure 2.1: A contact force is distributed across part of the surface of a rigid body. It is modeled by a force vector that acts in a point of application on the rigid body.

Postulate 2.1. A *force* is a vector quantity $\bar{F}$, which is assigned a *point of application* $\mathcal{P}$.

The effect of a force on a body is determined by the magnitude and the direction of the force and by its point of application. The force vector and the point of application define a line called the *line of action* (Fig. 2.1).

As is the case for all vectors, the force vector can be written as a sum of its components (Fig. 2.2),

$$\bar{F} = F_x \bar{e}_x + F_y \bar{e}_y + F_z \bar{e}_z, \tag{2.1}$$

or as a scalar F multiplied by a unit vector:

$$\bar{F} = F \bar{e}_F. \tag{2.2}$$

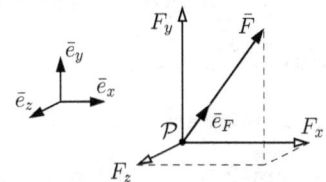

Figure 2.2: A force $\bar{F}$ acting in its point of application $\mathcal{P}$. Arrows with open arrowheads represent components of the force.

In Eq. (2.2), F is allowed to be negative, so that $F = \pm|\bar{F}|$. The force component of $\bar{F}$ in the λ direction is

$$F_\lambda = \bar{F} \cdot \bar{e}_\lambda = |\bar{F}| \cos\varphi, \tag{2.3}$$

where φ is the angle between $\bar{F}$ and $\bar{e}_\lambda$ (Fig. 2.3).

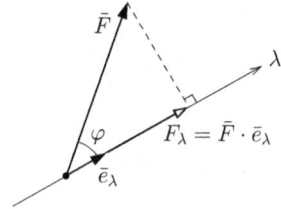

Figure 2.3: The force component of $\bar{F}$ w.r.t. the λ direction.

2.2 Moments and couples

Moments of force

To create a turning action about an axis, for instance when turning a bolt around its longitudinal axis, one applies a force in a point at some distance away from this axis (Fig. 2.4). The turning action of the force is called the *moment* of the force.

Definition 2.2 (Moment of force). Let $\bar{F}$ be a force with point of application $\mathcal{P}$. Then, the *moment* of the force $\bar{F}$, w.r.t. an arbitrary point $\mathcal{A}$, is the vector

$$\bar{M}_\mathcal{A} \equiv \overline{\mathcal{AP}} \times \bar{F}, \tag{2.4}$$

where $\mathcal{A}$ is the *moment reference point*.

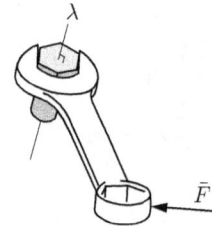

Figure 2.4: A force with its point of application at a distance from an axis λ creates a turning action about this axis.

Recollecting Def. A.18 of the cross product, the direction of the moment vector $\bar{M}_\mathcal{A}$ is given by the right-hand rule (Fig. 2.5). Therefore, the moment of force is perpendicular to the plane spanned by $\overline{\mathcal{AP}}$ and $\bar{F}$. The magnitude of $\bar{M}_\mathcal{A}$ is

$$\begin{aligned}|\bar{M}_\mathcal{A}| = |\overline{\mathcal{AP}} \times \bar{F}| &= \{\text{Eq. (A.19)}\} \\ &= |\overline{\mathcal{AP}}||\bar{F}|\sin\varphi \\ &= |\bar{F}|d_\perp, \end{aligned} \tag{2.5}$$

where $d_\perp = |\overline{\mathcal{AP}}|\sin\varphi$ is called the *lever arm* or the *moment arm*, and φ is the angle between $\overline{\mathcal{AP}}$ and $\bar{F}$ (Fig. 2.6). The moment vector is depicted as an arrow with a U-shaped head. The moment of force, w.r.t. an axis λ with direction vector $\bar{e}_\lambda$, is defined as

$$M_{\mathcal{A}\lambda} \equiv \bar{M}_\mathcal{A} \cdot \bar{e}_\lambda, \tag{2.6}$$

where $\mathcal{A}$ is an arbitrary point on the λ axis.

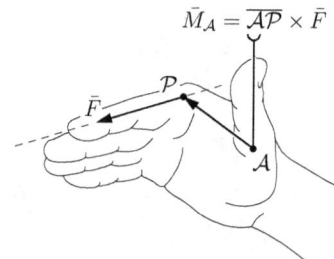

Figure 2.5: The right-hand rule for the moment of force. Keep the palm of your right hand aligned with the lever arm and angle your fingers in the direction of the force. Then your thumb will point in the direction of the moment vector.

Theorem 2.3. Let n forces, $\bar{F}_1, \ldots, \bar{F}_n$, act in the same point $\mathcal{P}$. The sum of their moments of force, w.r.t. an arbitrary point $\mathcal{A}$, is then equal to the moment of force of the sum of the force vectors, w.r.t. $\mathcal{A}$:

$$\sum_{i=1}^{n} \overline{\mathcal{AP}} \times \bar{F}_i = \overline{\mathcal{AP}} \times \sum_{i=1}^{n} \bar{F}_i. \tag{2.7}$$

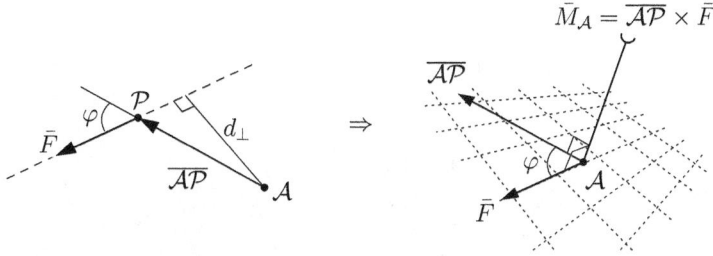

Figure 2.6: A force with force vector $\bar{F}$ and point of application $\mathcal{P}$ gives a moment of force $\bar{M}_{\mathcal{A}}$ w.r.t. $\mathcal{A}$, that is perpendicular to the plane spanned by $\overline{\mathcal{AP}}$ and $\bar{F}$.

Proof. The moment of force of the sum of the force vectors, w.r.t. $\mathcal{A}$, is

$$\overline{\mathcal{AP}} \times \sum_{i=1}^{n} \bar{F}_i = \overline{\mathcal{AP}} \times (\bar{F}_1 + \bar{F}_2 + \cdots + \bar{F}_n) = \{\text{Eq. (A.21b)}\}$$

$$= \overline{\mathcal{AP}} \times \bar{F}_1 + \overline{\mathcal{AP}} \times (\bar{F}_2 + \cdots + \bar{F}_n) = \{\text{repeat (A.21b)}\}$$

$$= \overline{\mathcal{AP}} \times \bar{F}_1 + \overline{\mathcal{AP}} \times \bar{F}_2 + \cdots + \overline{\mathcal{AP}} \times \bar{F}_n$$

$$= \sum_{i=1}^{n} \overline{\mathcal{AP}} \times \bar{F}_i. \qquad \square$$

When analyzing Statics problems, it is often convenient to decompose the forces into their vector components (Fig. 2.7). According to Theorem 2.3, the moment of a force is then given by the sum of the moments of force of its components.

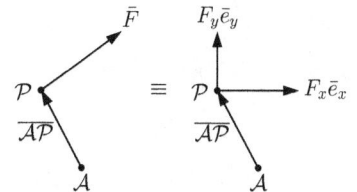

Figure 2.7: The moment of a force equals the sum of the moments of force of its vector components: $\overline{\mathcal{AP}} \times \bar{F} = \overline{\mathcal{AP}} \times F_x \bar{e}_x + \overline{\mathcal{AP}} \times F_y \bar{e}_y$ (2D).

Couples

Definition 2.4 (Force pair). A *force pair* consists of two forces, $\bar{F}$ with point of application $\mathcal{P}$ and $-\bar{F}$ with point of application $\mathcal{Q}$ (Fig. 2.8).

A trivial yet important property of the force pair is that the sum of its forces is $\bar{F} + (-\bar{F}) = \bar{0}$. Consequently, the only effect of a force pair is a turning action.

Definition 2.5 (Couple). A *couple* $\bar{C}$ is the sum of the moments of force from a force pair w.r.t. an arbitrary point $\mathcal{A}$.

Theorem 2.6. For an arbitrary force pair, $\bar{F}$ with point of application $\mathcal{P}$ and $-\bar{F}$ with point of application $\mathcal{Q}$ (Fig. 2.9), its couple is

$$\bar{C} = \overline{\mathcal{QP}} \times \bar{F}. \tag{2.8}$$

Proof. From Def. 2.5, it follows that the couple of a force pair, w.r.t. an arbitrary point $\mathcal{A}$, is

$$\bar{C} = \overline{\mathcal{AP}} \times \bar{F} + \overline{\mathcal{AQ}} \times (-\bar{F})$$

$$= \overline{\mathcal{AP}} \times \bar{F} - \overline{\mathcal{AQ}} \times \bar{F} = \{\text{Eq. (A.21b)}\}$$

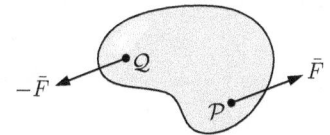

Figure 2.8: Illustration of a force pair.

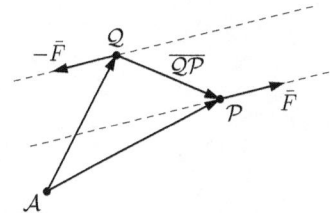

Figure 2.9: A force pair creating a couple $\bar{C} = \overline{\mathcal{QP}} \times \bar{F}$.

$$= (\overline{\mathcal{AP}} - \overline{\mathcal{AQ}}) \times \bar{F}$$
$$= (\overline{\mathcal{QA}} + \overline{\mathcal{AP}}) \times \bar{F} = \{\text{Parallelogram law}\}$$
$$= \overline{\mathcal{QP}} \times \bar{F}. \qquad\qquad\qquad \square$$

An example of a force pair is the action of a screwdriver on a slotted screw (Fig. 2.8). There are two contact points, $\mathcal{P}$ and $\mathcal{Q}$, between the screw head and the tip of the screwdriver, where two oppositely directed forces with the same magnitude act on the screw. The couple is independent of the choice of moment reference point. Therefore, this couple is a free vector that can be translated to an arbitrary point (Fig. 2.10).

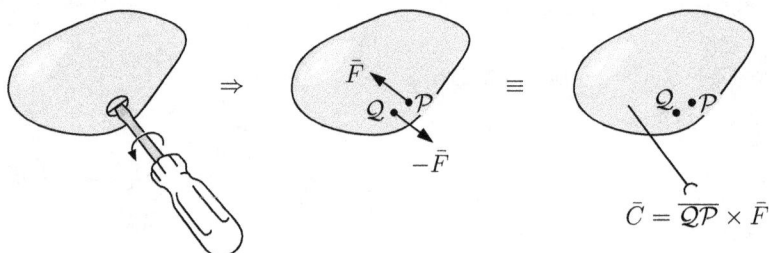

Figure 2.10: A screwdriver creates a turning action owing to the force pair acting on the slotted screw. The couple can be regarded as a free vector. That is, it does not act in any specific point on the rigid body.

2.3   Force-couple systems

Several forces and couples that act simultaneously on a rigid body form a forces-couple system.

Definition 2.7 (Force-couple system). A *force-couple system* Γ is a set of $n \geq 0$ forces $\bar{F}_1$, $\bar{F}_2, \ldots, \bar{F}_n$ with points of applications $\mathcal{P}_1$, $\mathcal{P}_2, \ldots, \mathcal{P}_n$, and a set of $m \geq 0$ couples $\bar{C}_1, \bar{C}_2, \ldots, \bar{C}_m$ (Fig. 2.11).

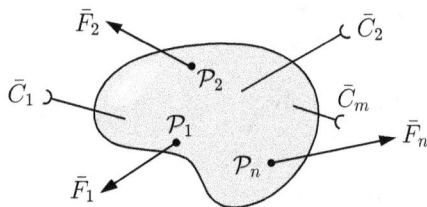

Figure 2.11: A force-couple system Γ with a number of forces and couples acting on a rigid body.

Force sums and moment sums

Definition 2.8 (Force sum). For a force-couple system, with notation as in Def. 2.7, the *force sum* is the vector

$$\Sigma \bar{F} \equiv \sum_{i=1}^{n} \bar{F}_i. \qquad\qquad (2.9)$$

Observe that the force sum is not assigned any point of application. Therefore, it does not fulfill the conditions in Postulate 2.1 for being a proper force.

Definition 2.9 (Moment sum). For a force-couple system Γ, with notation as in Def. 2.7, the *moment sum* w.r.t. an arbitrary point $\mathcal{A}$ is the vector

$$\Sigma \bar{M}_{\mathcal{A}} \equiv \sum_{i=1}^{n} \overline{\mathcal{AP}}_i \times \bar{F}_i + \sum_{j=1}^{m} \bar{C}_j. \tag{2.10}$$

Thus, one obtains the moment sum of a force-couple system, w.r.t. a point $\mathcal{A}$, by summing all the moments of force of the system w.r.t. $\mathcal{A}$, and all the couples of the system.

Theorem 2.10 (Transfer theorem of the moment sum). For a force-couple system, with notation as in Def. 2.7, and for two arbitrary points $\mathcal{A}$ and $\mathcal{B}$, it holds that

$$\Sigma \bar{M}_{\mathcal{B}} = \Sigma \bar{M}_{\mathcal{A}} + \overline{\mathcal{BA}} \times \Sigma \bar{F}, \tag{2.11}$$

where $\Sigma \bar{M}_{\mathcal{A}}$ and $\Sigma \bar{M}_{\mathcal{B}}$ are moment sums w.r.t. $\mathcal{A}$ and $\mathcal{B}$, respectively, and where $\Sigma \bar{F}$ is the force sum of the system.

Proof. Definition 2.9 gives

$$\Sigma \bar{M}_{\mathcal{B}} = \sum_{i=1}^{n} \overline{\mathcal{BP}}_i \times \bar{F}_i + \sum_{j=1}^{m} \bar{C}_j = \{\text{Parallelogram law}\}$$

$$= \sum_{i=1}^{n} \left(\overline{\mathcal{BA}} + \overline{\mathcal{AP}}_i \right) \times \bar{F}_i + \sum_{j=1}^{m} \bar{C}_j = \{\text{Eq. (A.21b)}\}$$

$$= \sum_{i=1}^{n} \overline{\mathcal{BA}} \times \bar{F}_i + \underbrace{\sum_{i=1}^{n} \overline{\mathcal{AP}}_i \times \bar{F}_i + \sum_{j=1}^{m} \bar{C}_j}_{= \Sigma \bar{M}_{\mathcal{A}}} = \{\text{Theorem 2.3}\}$$

$$= \overline{\mathcal{BA}} \times \Sigma \bar{F} + \Sigma \bar{M}_{\mathcal{A}}. \qquad \square$$

Reduced force-couple systems

Definition 2.11 (Reduced force-couple system). A *reduced force-couple system* $\Gamma_{\mathcal{A}}$ of a force-couple system Γ, w.r.t. a *point of reduction* $\mathcal{A}$, consists of the force sum $\Sigma \bar{F}$ of Γ acting in $\mathcal{A}$, and of a couple $\Sigma \bar{M}_{\mathcal{A}}$, being the moment sum of Γ w.r.t. $\mathcal{A}$ (Fig. 2.12).

The reduced force-couple system $\Gamma_{\mathcal{A}}$ is equivalent to Γ in the sense that Γ and $\Gamma_{\mathcal{A}}$ would induce the same motion for a rigid body.

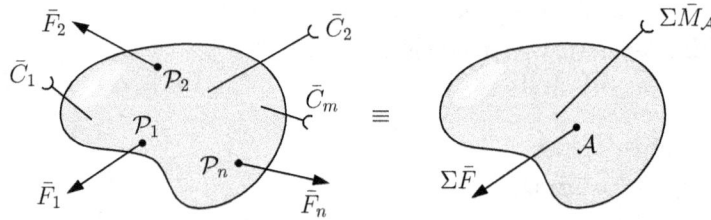

Figure 2.12: A force-couple system Γ, with an arbitrary number of forces and couples, is equivalent to its reduced force-couple system $\Gamma_{\mathcal{A}}$, w.r.t. an arbitrary point $\mathcal{A}$.

Definition 2.12 (Zero system)**.** If a force-couple system has the force sum $\Sigma\bar{F} = \bar{0}$ and the moment sum $\Sigma\bar{M}_{\mathcal{A}} = \bar{0}$, w.r.t. some point $\mathcal{A}$, then the force-couple system is a *zero system*.

Theorem 2.13. If a force-couple system is a zero system, then its moment sum is $\Sigma\bar{M}_{\mathcal{B}} = \bar{0}$ for every point $\mathcal{B}$.

Proof. When a force-couple system, with notation as in Def. 2.7, is a zero system, we have $\Sigma\bar{F} = \bar{0}$ and $\Sigma\bar{M}_{\mathcal{A}} = \bar{0}$ for some point $\mathcal{A}$. Consequently, by Theorem 2.10, it holds that

$$\Sigma\bar{M}_{\mathcal{B}} = \Sigma\bar{M}_{\mathcal{A}} + \overline{\mathcal{BA}} \times \Sigma\bar{F}$$
$$= \bar{0} + \overline{\mathcal{BA}} \times \bar{0}$$
$$= \bar{0}. \qquad \square$$

Theorem 2.13 states that a zero system is always a zero system, regardless of the choice of moment reference point.

2.4   Planar force-couple systems

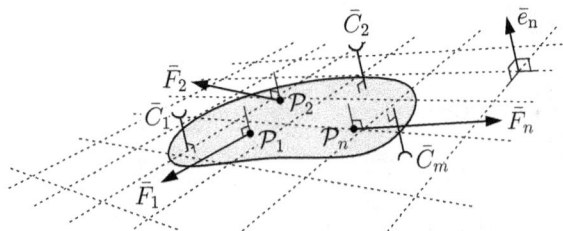

Figure 2.13: A planar force-couple system w.r.t. a reference plane with unit normal $\bar{e}_{\mathrm{n}}$.

Definition 2.14 (Planar force-couple system)**.** A force-couple system, with notation as in Def. 2.7, is *planar* if there exists a plane called the *reference plane*, such that each point of application $\mathcal{P}_i$ is located in this reference plane, and such that

$$\bar{F}_i \perp \bar{e}_{\mathrm{n}}, \quad i = 1, \ldots, n,$$
$$\bar{C}_j \parallel \bar{e}_{\mathrm{n}}, \quad j = 1, \ldots, m,$$

where $\bar{e}_{\mathrm{n}}$ is the unit normal of the reference plane (Fig. 2.13).

For a planar force-couple system and a moment reference point $\mathcal{A}$ located in its reference plane, the moments of force and the couples are oriented in the $\pm\bar{e}_n$ direction. Therefore, each moment of force and each couple can be uniquely represented by its normal component, which is a scalar. Figure 2.14 depicts a planar force-couple system with the xy plane as the reference plane. A scalar representation is used for the moments, which is indicated by a curved arrow for each couple $C_1, \ldots, C_m$, corresponding to the $\bar{e}_z$ or $-\bar{e}_z$ directions according to the right-hand rule (Fig. 2.15).

Let $\bar{F}$ denote a force, with point of application $\mathcal{P}$, belonging to a planar force-couple system (Fig. 2.16). The moment $\bar{M}_\mathcal{A} = \overline{\mathcal{AP}} \times \bar{F}$ of this force can be written as $\bar{M}_\mathcal{A} = M_\mathcal{A} \bar{e}_n$, where

$$
\begin{aligned}
M_\mathcal{A} &= \pm|\bar{M}_\mathcal{A}| = \{\text{Def. 2.2}\} \\
&= \pm|\overline{\mathcal{AP}} \times \bar{F}| = \{\text{Eq. (A.19)}\} \\
&= \pm|\overline{\mathcal{AP}}||\bar{F}|\sin\varphi.
\end{aligned}
$$

Here, φ is the angle between $\overline{\mathcal{AP}}$ and $\bar{F}$. Since the distance from $\mathcal{A}$ to the line of action of the force is $d_\perp = |\overline{\mathcal{AP}}|\sin\varphi$, it follows that

$$
M_\mathcal{A} = \pm F d_\perp. \tag{2.12}
$$

As pointed out above, the direction of the moment of force is dictated by the right-hand rule. The counterclockwise turning of the moment of force in Fig. 2.16 is oriented in the $\bar{e}_z$ direction. If we select the normal of the reference plane as $\bar{e}_n = \bar{e}_z$, the scalar representation $M_\mathcal{A}$ of this moment of force will be positive. Clockwise oriented moments of force, on the other hand, become negative. The converse applies if we select $\bar{e}_n = -\bar{e}_z$.

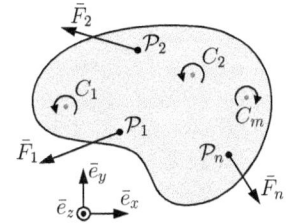

Figure 2.14: A planar force-couple system with the xy plane as its reference plane. The moments and couples of the system can be represented by scalars.

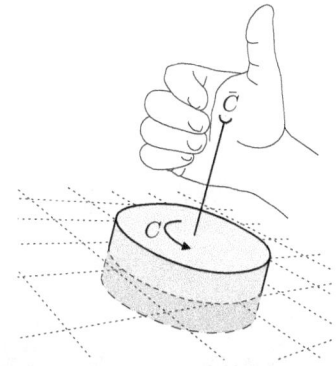

Figure 2.15: The vector direction of a couple represented by a scalar C and a curved arrow is identified by aligning the fingers of the right hand, except the thumb, with the curved arrow. The thumb then points in the vector direction of the couple.

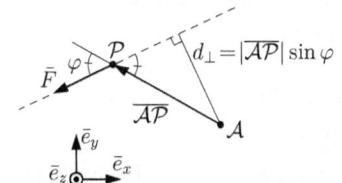

Figure 2.16: Geometry for the moment of the force F in a planar force-couple system, with the xy plane as its reference plane. The lever arm is denoted by $d_\perp$.

3

Static equilibrium

3.1 Equilibrium equations

Definition 3.1 (Static equilibrium). A body is in *static equilibrium* if every material point in the body has the same constant velocity relative to an inertial system.

Since Def. 3.1 requires that the velocities of the material points are equal and constant, it follows that all those points move along straight, parallel paths. Such motion is called *rectilinear translation* (Fig. 3.1). A rigid body is said to be at *rest*, if it is in static equilibrium and the velocity of the material points is zero in the chosen inertial system.

Static equilibrium is defined from the motion of the body, not from the forces that act on the body. If a rigid body is in static equilibrium, we need a postulate to identify force-couple systems that maintains this equilibrium:

Postulate 3.2 (Equilibrium conditions). A rigid body in static equilibrium remains in static equilibrium if the force-couple system that acts on this rigid body is a zero system,

$$\Sigma \bar{F} = \bar{0}, \tag{3.1a}$$

$$\Sigma \bar{M}_{\mathcal{A}} = \bar{0}, \tag{3.1b}$$

where $\Sigma \bar{F}$ is the force sum, and $\Sigma \bar{M}_{\mathcal{A}}$ is the moment sum of the force-couple system w.r.t. an arbitrary point $\mathcal{A}$.

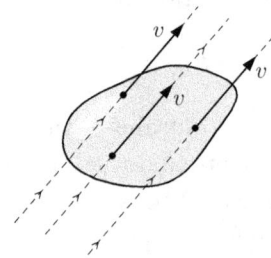

Figure 3.1: Static equilibrium implies that a rigid body describes rectilinear translation. Every point in the body moves with the same constant velocity.

Equation (3.1a) is called the *Force equilibrium equation* and Eq. (3.1b) is called the *Moment equilibrium equation*. According to Theorem 2.13, the moment reference point of the Moment equilibrium equation can be freely selected.

The Force and the Moment equilibrium equations are vector equations. According to Eq. (A.11) they can be written on component form as a system of six scalar equations:

$$\begin{cases} \Sigma F_x = 0, & \Sigma M_{Ax} = 0, \\ \Sigma F_y = 0, & \Sigma M_{Ay} = 0, \\ \Sigma F_z = 0, & \Sigma M_{Az} = 0. \end{cases}$$

Equilibrium of planar force-couple systems

For a planar force-couple system, the Force and the Moment equilibrium equations can be simplified by choosing a coordinate system with two coordinate axes in the reference plane. If the xy plane is placed in the reference plane (Fig. 2.14), so that $\bar{e}_n = \bar{e}_z$, then Def. 2.14 gives:

$$\bar{F}_i \perp \bar{e}_z \quad \Leftrightarrow \quad F_{iz} = 0, \qquad i = 1, 2, \ldots \qquad \Rightarrow$$
$$\Sigma F_z = 0.$$

Moreover, all the moments of force and all the couples are oriented in the z direction (Sect. 2.4), so that

$$\Sigma M_{Ax} = \Sigma M_{Ay} = 0,$$

where $\mathcal{A}$ is a moment reference point in the reference plane. In conclusion, only three nontrivial, scalar equilibrium equations remain for the planar force-couple system:

$$\begin{cases} \Sigma F_x = 0, \\ \Sigma F_y = 0, \\ \Sigma M_{Az} = 0. \end{cases}$$

3.2   Free-body diagrams

A *free-body diagram* is an aid to identify all forces and couples that act on a mechanical system. When drawing a free-body diagram, the body is isolated from its surroundings. That is, the surrounding objects are removed and instead their action on the body is represented by forces and couples. When drawing a free-body diagram, one observes the following protocol:

1. Decide which body to analyze. Here, a dashed line circumscribes the selected body.

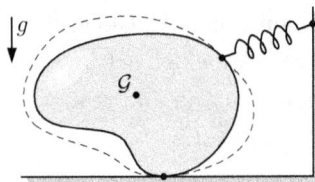

2. Draw a diagram containing *only*
the selected body.

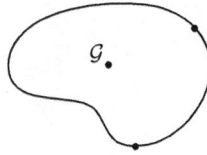

3. Represent the action of the surrounding objects on the body with
forces and couples.

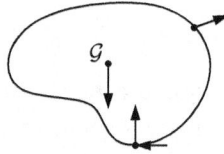

The surroundings of the analyzed body may exert forces through force
fields, *e.g.* the force of gravity, as well as contact forces that arise at points
of physical contact between the body and the surrounding objects.

The force of gravity

The action of *gravity* on a rigid body near the surface of the Earth is
modeled as a force: the *force of gravity* acting in the center of gravity $\mathcal{G}$
of a body (Fig. 3.2). The force of gravity is directed towards the center
of the Earth, and it has the magnitude mg, where m is the mass of the
body, and g is the local gravity constant. Gravity will be further studied
in Chapter 4.

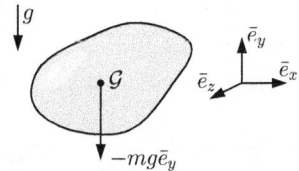

Figure 3.2: Gravity acting on a rigid
body near the surface of the Earth. The
force of gravity has the magnitude mg,
acting in the center of gravity $\mathcal{G}$ of the
rigid body.

Constraint forces and couples

If a rigid body is in physical contact with surrounding objects, and therefore is prevented from moving or rotating freely, then *constraint forces*
and *constraint couples* arise at the points of contact.

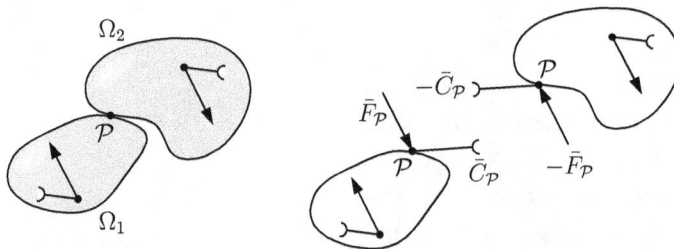

Figure 3.3: Two bodies, Ω_1 and Ω_2, with
a point contact $\mathcal{P}$. The free-body diagrams show contact forces and couples
between the bodies.

Consider a *point contact* between two bodies, Ω_1 and Ω_2, that are in
physical contact with each other at a common point $\mathcal{P}$. Generally, this
sort of contact creates a couple $\bar{C}_{\mathcal{P}}$ and a force $\bar{F}_{\mathcal{P}}$, which act in $\mathcal{P}$ on Ω_1.
According to an extended version of the Law of action and reaction, the
contact also creates a couple $-\bar{C}_{\mathcal{P}}$ and a force $-\bar{F}_{\mathcal{P}}$, which act in $\mathcal{P}$
on Ω_2 (Fig. 3.3).

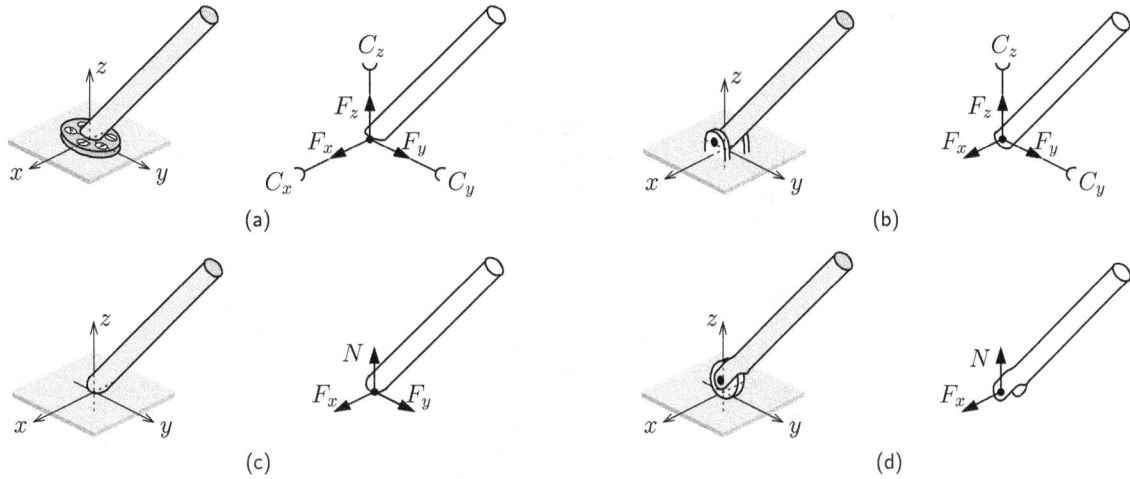

(a)

(b)

(c)

(d)

Figure 3.4: Free-body diagrams for different supports. (a) Rigid supports, for instance welds, bolted joints and glued joints, where forces and couples can arise in any direction. (b) A hinge where a pin allows rotation around the x axis. As a consequence, $C_x = 0$. (c) At a friction contact of a rounded body, rotations are allowed through rolling against the surface: $C_x = C_y = 0$. By neglecting friction for rotation around the normal axis, we obtain $C_z = 0$. (d) A wheel eliminates one of the friction force components, $F_y = 0$, while rotation is allowed around all axes: $C_x = C_y = C_z = 0$.

The point contact model is used for different types of supports and joints between bodies, such as welds, hinges, bearings and so on. The type of joint determines the direction of the constraint force and couple according to the two following principles:

1. If a joint at $\mathcal{P}$ allows for Ω_1 to translate freely relative to Ω_2 in a direction $\bar{e}_\lambda$, then

$$\bar{F}_{\mathcal{P}} \cdot \bar{e}_\lambda = 0.$$

An example is the y direction in Fig. 3.4d, where $F_y = 0$.

2. If a joint at $\mathcal{P}$ allows for Ω_1 to rotate freely relative to Ω_2 around an axis through $\mathcal{P}$ with direction vector $\bar{e}_\lambda$, then

$$\bar{C}_{\mathcal{P}} \cdot \bar{e}_\lambda = 0.$$

An example is the x direction in Fig. 3.4b where $C_x = 0$.

Thus, constraint forces can only arise in directions where relative displacement is restricted. Similarly, constraint couples can only arise in directions where relative rotation is restricted.

There are many different types of supports. Therefore, one needs to formulate a suitable point contact model for every new case. Some examples are given in Fig. 3.4. When a new type of support is encountered, first assume that all the components of the constraint force and couple are non-zero. From that outset, methodically proceed to eliminate components that lack constraint.

Strings and pulleys

A *string* is an idealized rope, wire or the like, which can be regarded as inextensible and massless. A stretched string is loaded only by a tensile force $T > 0$ in the longitudinal direction of the string. Thus, the ends of a segment of a stretched string are loaded by two forces $\bar{T}$ and $-\bar{T}$, respectively, parallel to the string (Fig. 3.5a).

When a string runs over a frictionless *pulley* with negligible mass, the tensile force will be the same in both ends of the string. This becomes clear from the moment equilibrium w.r.t. the hub of the pulley (Fig. 3.5b).

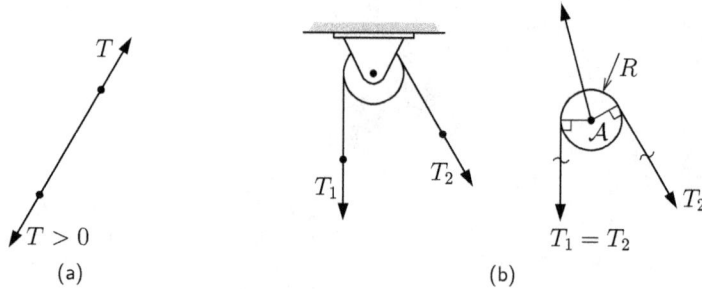

(a)   (b)

Figure 3.5: (a) A stretched string is loaded by two oppositely directed forces, which are parallel to the string. (b) A string runs over a frictionless pulley. A moment equilibrium for the pulley w.r.t. $\mathcal{A}$ gives $RT_1 - RT_2 = 0$, showing that $T_1 = T_2$.

Two-force members

An important special case of equilibrium is when precisely two forces act on a rigid body, which is then called a *two-force member*.

Theorem 3.3 (Two-force member). If precisely two nonzero forces, and no couple, act on a rigid body in static equilibrium, then those forces have the same magnitude, are oppositely directed and have coincident lines of action (Fig. 3.6).

Proof. Let two arbitrary forces, $\bar{F}_{\mathcal{P}}$ with point of application $\mathcal{P}$ and $\bar{F}_{\mathcal{Q}}$ with point of application $\mathcal{Q}$, act on a rigid body in static equilibrium. The force equilibrium gives

$$\bar{F}_{\mathcal{P}} + \bar{F}_{\mathcal{Q}} = \bar{0},$$

so that $\bar{F}_{\mathcal{P}} = -\bar{F}_{\mathcal{Q}}$, showing that the forces have the same magnitude, and that they are oppositely directed. Thus, their lines of action are parallel.

A moment equilibrium w.r.t. $\mathcal{P}$ gives (Fig. 3.7)

$$\overline{\mathcal{PQ}} \times \bar{F}_{\mathcal{Q}} = \bar{0} \quad \Leftrightarrow \quad \{\text{Eq. (A.19)}\} \quad \Leftrightarrow$$
$$|\overline{\mathcal{PQ}}||\bar{F}_{\mathcal{Q}}| \sin \varphi = 0 \quad \Leftrightarrow \quad \{\bar{F}_{\mathcal{Q}} \neq \bar{0}\} \quad \Leftrightarrow$$
$$|\overline{\mathcal{PQ}}| \sin \varphi = 0,$$

where φ is the angle between $\overline{\mathcal{PQ}}$ and $\bar{F}_{\mathcal{Q}}$. The perpendicular distance between the lines of action is (Fig. 3.7)

$$d_{\perp} = |\overline{\mathcal{PQ}}| \sin \varphi = 0.$$

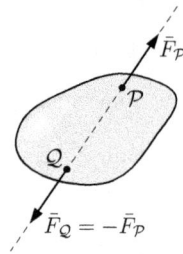

Figure 3.6: A two-force member in static equilibrium, where the lines of action of the forces are coincident.

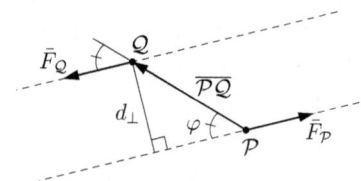

Figure 3.7: Geometry for the proof of Theorem 3.3.

Therefore, the lines of action must coincide.   □

Massless bars attached to hinges are prototypical two-force members (Fig. 3.8). The analysis of multi-body mechanical systems can sometimes be simplified considerably by exploiting this property.

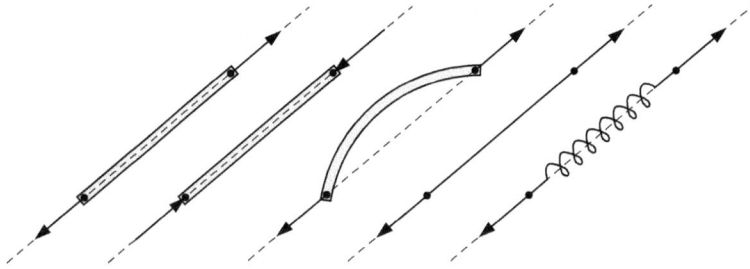

Figure 3.8: Bars, strings and springs with negligible mass, free from couples at their joints, are two-force members.

3.3   Multi-body mechanical systems

When a construction contains several different members, that are all in static equilibrium, the force-couple system of each member must be a zero system. It can be shown that it is a necessary condition for static equilibrium of a multi-body system, that a zero system of external forces and couples acts on the mechanical system.

When analyzing a multi-body system, it is permissible to draw the free-body diagram comprising several connected rigid bodies. For instance, consider the excavator in Fig. 3.9a. Depending on the problem formulation, it can be convenient to either draw the free-body diagram for the entire excavator (Fig. 3.9b), or to draw the free-body diagram for each of its members (Fig. 3.9c). The latter alternative is more suitable if the problem formulation concerns internal forces between the members of the construction.

The free-body diagrams of the members of the excavator (Fig. 3.9c) show some useful principles: Forces and reaction forces arise in the contact points between each pair of members. According to the Law of action and reaction, the force and the reaction force have the same magnitude, but opposite directions. The hydraulic cylinder is assumed to be massless, and therefore it can be regarded as a two-force member. Thus, the forces that act in its ends have the same magnitude, are oppositely directed, and their lines of action coincide (Theorem 3.3). The equilibrium equations can be formulated for each member, or for the entire mechanical system.

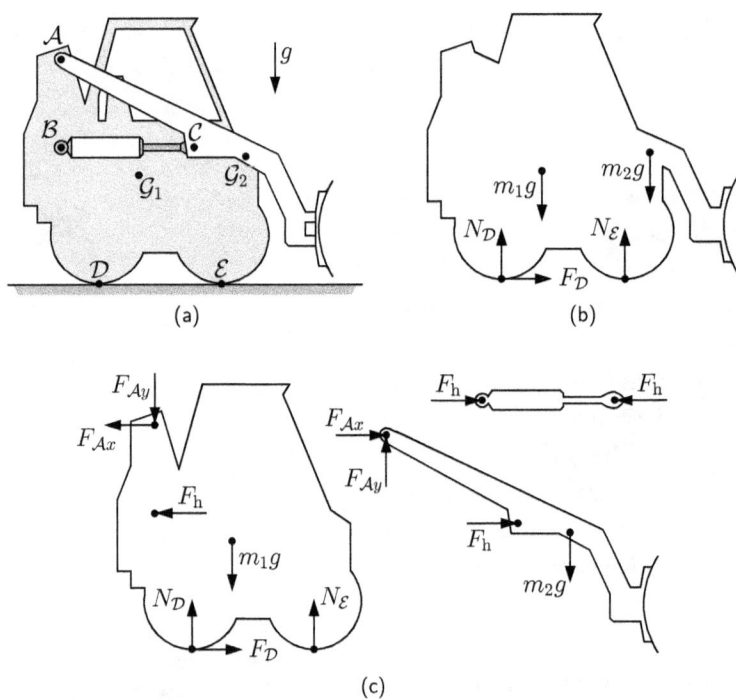

Figure 3.9: (a) An excavator consisting of a vehicle, with center of gravity $\mathcal{G}_1$ and mass m_1, a massless hydraulic cylinder $\mathcal{BC}$, and a scraper blade arm with center of gravity $\mathcal{G}_2$ and mass m_2. The front wheels are unbraked. (b) A free-body diagram of the whole construction. (c) Free-body diagrams of the members of the mechanical system, where the hydraulic cylinder is regarded as a two-force member.

4
Center of mass

4.1 Density

The *density* ϱ of a material is defined as its mass per unit volume, with the SI units of kg/m^3 and the USC units of slug/ft^3. Since a body may comprise several different materials, the density of the body can vary in space: $\varrho = \varrho(\bar{r})$. The mass of a body Ω is therefore

$$m = \int_\Omega \mathrm{d}m = \int_\Omega \varrho(\bar{r})\mathrm{d}V, \qquad (4.1)$$

where $\mathrm{d}V$ is an infinitesimal volume element, $\mathrm{d}m = \varrho\mathrm{d}V$ is a mass element, and $\bar{r}$ is the position vector of this mass element (Fig. 4.1).

4.2 Center of mass

Consider a rigid body near the surface of the Earth. If the body is suspended by a string connected to a point $\mathcal{P}_1$ on the surface of the body, then this string defines a vertical line through the body. If this procedure is repeated for several points, $\mathcal{P}_1, \mathcal{P}_2, \ldots$, it is an experimental fact that all the corresponding vertical lines intersect at a common, body-fixed point called the *center of gravity* (Fig. 4.2).

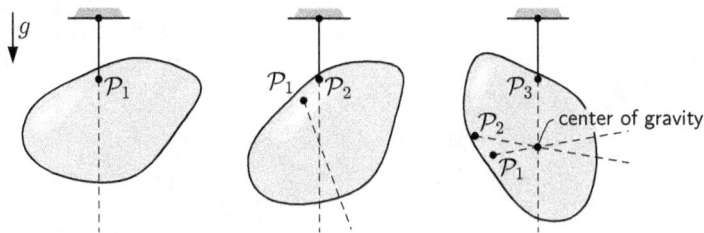

Figure 4.1: Geometry for the integral expression for the mass of a body.

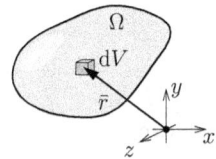

Figure 4.2: The vertical lines for different points of suspension $\mathcal{P}_1, \mathcal{P}_2, \ldots$ on the body intersect a common point, which is called the center of gravity.

First, a definition of the center of mass $\mathcal{G}$ of a body is given. Later, it will be shown that the center of gravity always coincides with the center of mass.

Definition 4.1 (Center of mass). For a body Ω with mass m and density $\varrho(\bar{r})$, the *center of mass* $\mathcal{G}$ is defined by its position vector

$$\bar{r}_\mathcal{G} \equiv \frac{1}{m} \int_\Omega \bar{r} \mathrm{d}m = \frac{1}{m} \int_\Omega \bar{r} \varrho(\bar{r}) \mathrm{d}V. \tag{4.2}$$

This means that if $\bar{r}_\mathcal{G} = x_\mathcal{G} \bar{e}_x + y_\mathcal{G} \bar{e}_y + z_\mathcal{G} \bar{e}_z$, then the x coordinate of the center of mass is given by

$$x_\mathcal{G} = \frac{1}{m} \int_\Omega x \varrho(x, y, z) \mathrm{d}x \mathrm{d}y \mathrm{d}z, \tag{4.3}$$

with analogous expressions for $y_\mathcal{G}$ and $z_\mathcal{G}$.

Theorem 4.2 (Center of mass of a composite body). If a body Ω with mass m is composed of n parts $\Omega_1, \ldots, \Omega_n$, then the center of mass of the composite body is

$$\bar{r}_\mathcal{G} = \frac{1}{m} \sum_{i=1}^n m_i \bar{r}_{\mathcal{G}i}, \tag{4.4}$$

where m_i is the mass of Ω_i, and $\bar{r}_{\mathcal{G}i}$ is the position vector of the center of mass of Ω_i (Fig. 4.3).

Proof. According to Def. 4.1 concerning the center of mass, we have

$$\bar{r}_\mathcal{G} = \frac{1}{m} \int_\Omega \bar{r} \mathrm{d}m = \{\text{one integral for each part}\}$$

$$= \frac{1}{m} \left(\int_{\Omega_1} \bar{r} \mathrm{d}m + \cdots + \int_{\Omega_n} \bar{r} \mathrm{d}m \right)$$

$$= \frac{1}{m} \left(m_1 \underbrace{\frac{1}{m_1} \int_{\Omega_1} \bar{r} \mathrm{d}m}_{=\bar{r}_{\mathcal{G}1}} + \cdots + m_n \underbrace{\frac{1}{m_n} \int_{\Omega_n} \bar{r} \mathrm{d}m}_{=\bar{r}_{\mathcal{G}n}} \right)$$

$$= \frac{1}{m} \sum_{i=1}^n m_i \bar{r}_{\mathcal{G}i}. \qquad \square$$

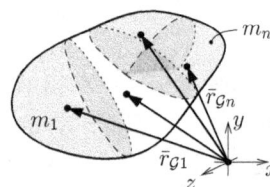

Figure 4.3: A body composed of several parts Ω_i, $i = 1, \ldots, n$, with masses m_i and centers of mass $\mathcal{G}_i$.

Definition 4.3 (Centroid). The *centroid* $\mathcal{C}$ of a body Ω is defined by its position vector

$$\bar{r}_\mathcal{C} \equiv \frac{1}{V} \int_\Omega \bar{r} \mathrm{d}V, \tag{4.5}$$

where $V = \int_\Omega \mathrm{d}V$ is the volume of the body.

It is common that an entire body Ω consists of a single type of material, with a constant density ϱ across Ω. Such a body is said to be *homogeneous*, and its mass is $m = \varrho V$. Therefore, according to Eq. (4.2), its center of mass is

$$\bar{r}_\mathcal{G} = \frac{1}{m} \int_\Omega \bar{r} \varrho \mathrm{d}V = \frac{1}{\varrho V} \varrho \int_\Omega \bar{r} \mathrm{d}V = \frac{1}{V} \int_\Omega \bar{r} \mathrm{d}V = \bar{r}_\mathcal{C}.$$

Thus, for homogeneous rigid bodies, the center of mass and the centroid are coincident.

4.3 Center of mass of slender bodies

For a thin shell represented by a surface Π in space, the *surface density* ϱ_A is defined as the mass per unit area of the shell with units of kg/m^2 or slug/ft^2. Since this surface density can vary across the shell, we write $\varrho_A = \varrho_A(\bar{r})$, $\bar{r} \in \Pi$. Let $\mathrm{d}A$ denote an infinitesimal surface element on Π (Fig. 4.4). Then, the corresponding mass element is $\mathrm{d}m = \varrho_A \mathrm{d}A$, so that the position vector for the center of mass $\mathcal{G}$ of the shell is

$$\bar{r}_{\mathcal{G}} = \frac{1}{m} \int_{\Pi} \bar{r} \mathrm{d}m = \frac{1}{m} \int_{\Pi} \bar{r} \varrho_A \mathrm{d}A, \qquad (4.6)$$

according to Eq. (4.2). In the same way, Eq. (4.5) gives the centroid of the shell:

$$\bar{r}_C = \frac{1}{A} \int_{\Pi} \bar{r} \mathrm{d}A, \qquad (4.7)$$

where $A = \int_{\Pi} \mathrm{d}A$ is the area of the shell.

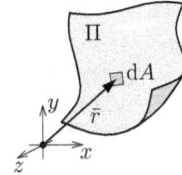

Figure 4.4: Geometry for the definition of the center of mass of a thin shell Π.

For a slender bar, whose backbone follows the curve Λ from $\mathcal{P}$ to $\mathcal{Q}$, the *line density* ϱ_ℓ is defined as the mass per unit length of the bar with units of kg/m or slug/ft. Let $\mathrm{d}s$ denote an infinitesimal line element on Λ, so that the corresponding mass element is $\mathrm{d}m = \varrho_\ell \mathrm{d}s$ (Fig. 4.5). Then, according to Eq. (4.2), the center of mass $\mathcal{G}$ of the bar is

$$\bar{r}_{\mathcal{G}} = \frac{1}{m} \int_{\Lambda} \bar{r} \mathrm{d}m = \frac{1}{m} \int_{\Lambda} \bar{r} \varrho_\ell \mathrm{d}s. \qquad (4.8)$$

Moreover, in the case of a slender bar, Eq. (4.5) becomes

$$\bar{r}_C = \frac{1}{\ell} \int_{\Lambda} \bar{r} \mathrm{d}s, \qquad (4.9)$$

where $\ell = \int_{\Lambda} \mathrm{d}s$ is the arc length of Λ.

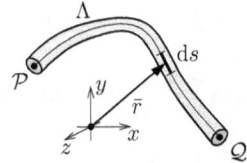

Figure 4.5: Geometry for the definition of the center of mass of a slender bar along a curve Λ.

4.4 Center of gravity

Gravity is a body force that acts across the entire region occupied by the body. Near the surface of the Earth, a body with density $\varrho = \varrho(\bar{r})$ is affected by a body force $\varrho(\bar{r})\bar{g}$, where $\bar{g}$ denotes the constant, downward field of gravity with magnitude g.

Theorem 4.4 (Center of gravity). For a body with mass m and density $\varrho = \varrho(\bar{r})$ in a constant field of gravity $\bar{g}$, the force sum of the body force $\varrho(\bar{r})\bar{g}$ of gravity is

$$\bar{F}_{\mathrm{g}} = m\bar{g}, \qquad (4.10)$$

and the moment sum of $\varrho(\bar{r})\bar{g}$ w.r.t. the center of mass $\mathcal{G}$ of the body is $\Sigma \bar{M}_{\mathcal{G}} = \bar{0}$.

Proof. Consider an arbitrary volume element $\mathrm{d}V$ with mass $\mathrm{d}m = \varrho\mathrm{d}V$ and position vector $\bar{r}$. The force of gravity on this volume element is $\mathrm{d}\bar{F} = \bar{g}\mathrm{d}m$ (Fig. 4.6), so that the force sum acting on the body Ω is

$$
\begin{aligned}
\bar{F}_{\mathrm{g}} &= \int_\Omega \mathrm{d}\bar{F} \\
&= \int_\Omega \bar{g}\mathrm{d}m = \{\bar{g}\ \text{const.}\} \\
&= \bar{g}\int_\Omega \mathrm{d}m = \{\text{Eq. (4.1)}\} \\
&= m\bar{g}.
\end{aligned}
$$

The volume elements are regarded as particles (Postulate 1.2). Therefore, the couple acting on $\mathrm{d}V$ is assumed to be $\bar{0}$. The moment sum w.r.t. $\mathcal{G}$ is

$$
\begin{aligned}
\Sigma\bar{M}_{\mathcal{G}} &= \int_\Omega (\bar{r} - \bar{r}_{\mathcal{G}}) \times \mathrm{d}\bar{F} \\
&= \int_\Omega (\bar{r} - \bar{r}_{\mathcal{G}}) \times \bar{g}\mathrm{d}m = \{\bar{g}\ \text{const.}\} \\
&= \left[\int_\Omega (\bar{r} - \bar{r}_{\mathcal{G}})\mathrm{d}m\right] \times \bar{g} \\
&= \left(\int_\Omega \bar{r}\mathrm{d}m - \int_\Omega \bar{r}_{\mathcal{G}}\mathrm{d}m\right) \times \bar{g} = \{\bar{r}_{\mathcal{G}}\ \text{const.}\} \\
&= \left(m\frac{1}{m}\int_\Omega \bar{r}\mathrm{d}m - \bar{r}_{\mathcal{G}}\int_\Omega \mathrm{d}m\right) \times \bar{g} = \{\text{Def. 4.1 and Eq. (4.1)}\} \\
&= (m\bar{r}_{\mathcal{G}} - m\bar{r}_{\mathcal{G}}) \times \bar{g} \\
&= \bar{0}. \qquad\qquad\qquad\qquad\qquad\qquad\qquad\qquad\qquad\qquad\square
\end{aligned}
$$

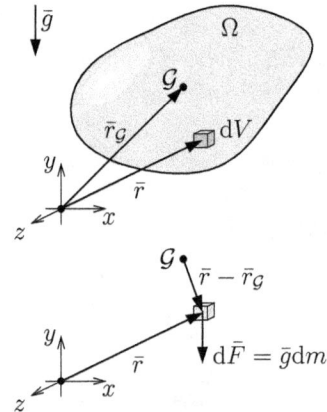

Figure 4.6: Geometry for the action of gravity on a rigid body, and for the action of gravity on a volume element.

According to Theorem 4.4, gravity can be reduced to a single force $m\bar{g}$ acting in the center of mass $\mathcal{G}$ of the body. Again, consider the suspended body in Fig. 4.2. Only two forces act on this body: the force of the string and that of gravity (Fig. 4.7). Thus, the body becomes a two-force member. Therefore, according to Theorem 3.3, $\mathcal{G}$ is located on the vertical line corresponding to any possible point of suspension. Hence, the center of gravity is coincident with the center of mass.

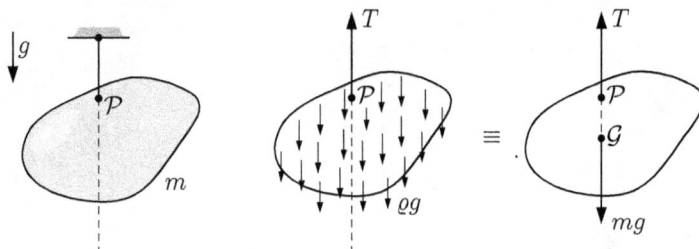

Figure 4.7: The field of gravity $\bar{g}$ acts on a rigid body that is suspended near the surface of the Earth. The force field is reduced to one force with magnitude mg and its point of application in the center of mass $\mathcal{G}$.

5

Distributed and internal forces

5.1   Surface and line loads

On a surface Π with unit normal $\bar{e}_n$, the normally-directed surface load is
a vector-valued function $\bar{f}(\bar{r}) = f(\bar{r})\bar{e}_n$ with the units of N/m^2 or lb$_f$/ft^2.
Here, $\bar{r} \in \Pi$ denotes the position vector (Fig. 5.1). For simplicity, we
restrict our attention to planar surfaces.

Definition 5.1 (Resultant and center of loading). For a planar surface Π
with unit normal $\bar{e}_n$, the *resultant* of the normally-directed surface
load $f(\bar{r})\bar{e}_n$ is a force

$$\bar{F} = F\bar{e}_n, \qquad F = \int_\Pi f(\bar{r})\mathrm{d}A, \tag{5.1}$$

and its point of application is the *center of loading* $\mathcal{C}$, with position
vector

$$\bar{r}_\mathcal{C} \equiv \frac{1}{F} \int_\Pi \bar{r}f(\bar{r})\mathrm{d}A, \tag{5.2}$$

when $F \neq 0$. Otherwise, $\mathcal{C}$ is undefined.

The force vector $\bar{F}$ of the resultant is the force sum of the surface load,
and the center of loading $\mathcal{C}$ is defined so that the moment sum of the
surface load w.r.t. $\mathcal{C}$ vanishes:

Theorem 5.2. For a planar surface Π with unit normal $\bar{e}_n$, and a surface
load $f(\bar{r})\bar{e}_n$, the moment sum of the surface load, w.r.t. its center of
loading $\mathcal{C}$, is $\Sigma\bar{M}_\mathcal{C} = \bar{0}$.

Proof. Consider a surface element $\mathrm{d}A$ with position vector $\bar{r}$, subjected
to the load $\mathrm{d}\bar{F} = f(\bar{r})\bar{e}_n\mathrm{d}A$ (Fig. 5.2). The couple on each surface
element is assumed to be $\bar{0}$. Thus, the moment sum w.r.t. $\mathcal{C}$ is

$$\Sigma\bar{M}_\mathcal{C} = \int_\Pi (\bar{r} - \bar{r}_\mathcal{C}) \times \mathrm{d}\bar{F} = \{\bar{e}_n \text{ const.}\}$$

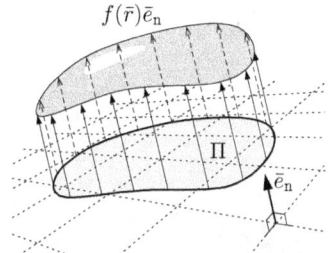

Figure 5.1: A normally-directed surface
load $\bar{f}(\bar{r}) = f(\bar{r})\bar{e}_n$, $\bar{r} \in \Pi$, acting on a
planar surface Π.

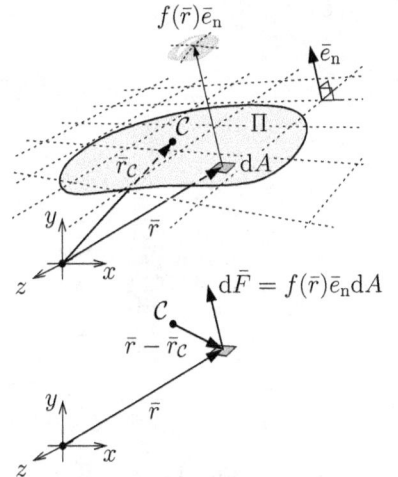

Figure 5.2: Geometry for Theorem 5.2.

$$= \left[\int_\Pi (\bar{r} - \bar{r}_C) f(\bar{r}) \mathrm{d}A \right] \times \bar{e}_\mathrm{n}$$

$$= \left[\int_\Pi \bar{r} f(\bar{r}) \mathrm{d}A - \int_\Pi \bar{r}_C f(\bar{r}) \mathrm{d}A \right] \times \bar{e}_\mathrm{n} = \{ \bar{r}_C \text{ const.} \}$$

$$= \left[F \underbrace{\frac{1}{F} \int_\Pi \bar{r} f(\bar{r}) \mathrm{d}A}_{=\bar{r}_C} - \bar{r}_C \underbrace{\int_\Pi f(\bar{r}) \mathrm{d}A}_{=F} \right] \times \bar{e}_\mathrm{n}$$

$$= (F\bar{r}_C - F\bar{r}_C) \times \bar{e}_\mathrm{n}$$

$$= \bar{0}. \qquad\qquad \square$$

Theorem 5.2 asserts that the normally-directed surface load can be reduced to one force, the resultant $\bar{F}$ acting in the center of loading $\mathcal{C}$ (Fig. 5.3). This underscores the general applicability of the concept of a *force*, as introduced in Postulate 2.1.

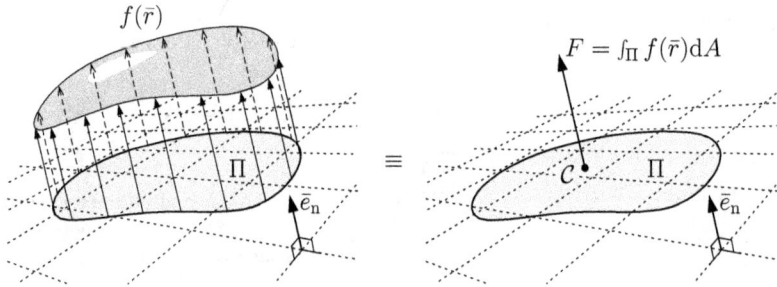

Figure 5.3: A surface load on a planar surface can be represented by one force with its point of application in the plane of this surface.

Choose a rectangular coordinate system xyz so that the loaded region Π is in the xy plane. If $\Pi = [x_0, x_1] \times [-b/2, b/2]$, that is, if Π is a strip with width b, then we can interpret the normally-directed surface load $f(x, y)$ as a *line load*,

$$q(x) = \int_{-b/2}^{b/2} f(x, y) \mathrm{d}y, \qquad x \in [x_0, x_1], \tag{5.3}$$

with the units of N/m or $\mathrm{lb_f}$/ft (Fig. 5.4).

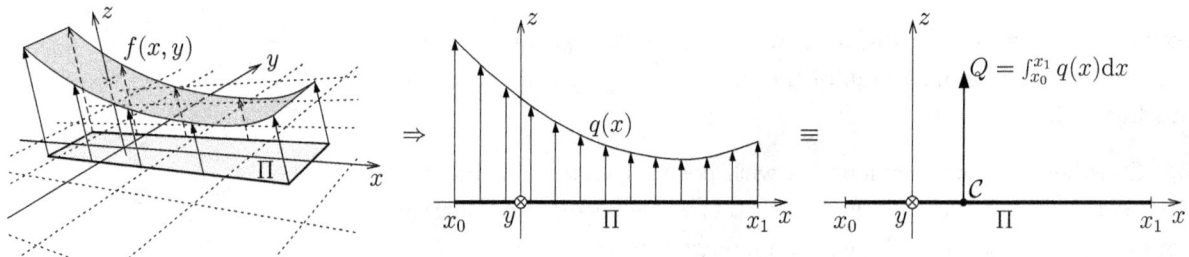

The resultant $\bar{Q} = Q\bar{e}_z$ of the line load $q(x)$ is given by Eq. (5.1), and

Figure 5.4: A surface load f acting on a strip Π can be represented by a line load q, which, in turn, can be represented by the resultant Q, with its point of application in the center of loading $\mathcal{C}$.

its magnitude becomes

$$Q = \int_{x_0}^{x_1} q(x)\mathrm{d}x. \tag{5.4}$$

The position of the center of loading on the x axis follows from Eq. (5.2):

$$x_C = \frac{1}{Q} \int_{x_0}^{x_1} xq(x)\mathrm{d}x, \qquad Q \neq 0. \tag{5.5}$$

Observe that lower-case letters denote distributed loads, *e.g.* f and q, while upper-case letters denote the corresponding resultants F and Q, with the units of force.

If $q(x) \geq 0$, $\forall x$, or if $q(x) \leq 0$, $\forall x$, then the center of loading on the x axis is the same as the centroid of the surface under the graph of $q(x)$ (Fig. 5.5).

5.2   Internal forces and couples

Definition 5.3 (Planar section). A *planar section* through a body Ω is a planar surface Π_λ, with normal direction $\bar{e}_\lambda$, in the interior of Ω (Fig. 5.6a).

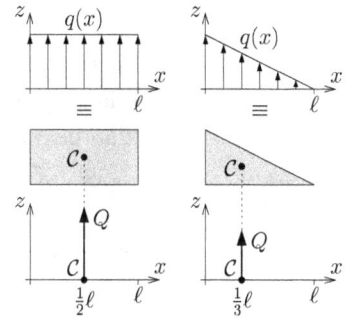

Figure 5.5: The position of the center of loading is the same as the centroid of the surface below the graph of $q(x)$.

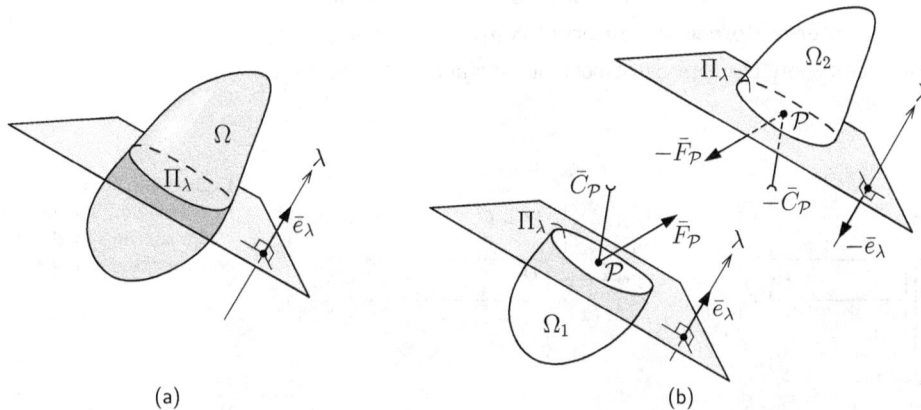

(a)   (b)

Figure 5.6:   (a) A planar section Π_λ through a body Ω. (b) Free-body diagrams showing the parts of the body created by the section Π_λ. The interaction between the parts is modeled as a point contact.

To analyze the internal forces of a body Ω, we place a planar section Π_λ through Ω, which divides Ω into two parts: Ω_1 with outward normal $\bar{e}_\lambda$ on Π_λ, and Ω_2 with outward normal $-\bar{e}_\lambda$ on Π_λ. Moreover, the interaction between the two parts of the body is modeled as a point contact at a point $\mathcal{P}$ in Π_λ. A *section force* $\bar{F}_\mathcal{P}$ and a *section couple* $\bar{C}_\mathcal{P}$ act in $\mathcal{P}$ on Ω_1. Conversely, $-\bar{F}_\mathcal{P}$ and $-\bar{C}_\mathcal{P}$ act in $\mathcal{P}$ on Ω_2 (Fig. 5.6b).

Henceforth, we will consider straight, slender bodies called *beams*. We introduce a rectangular coordinate system xyz, where the x direction is aligned with the longitudinal direction of the beam. The shape of

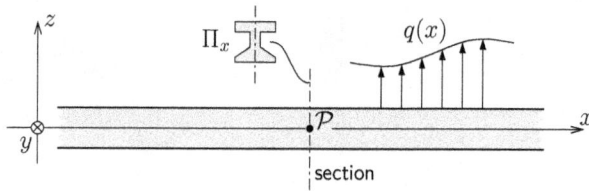

Figure 5.7: Geometry and loading of a beam. The cross-section of the beam Π_x has mirror symmetry w.r.t. the xz plane.

the beam cross-section Π_x, with normal direction $\bar{e}_x$, is assumed to be unchanging along the length of the beam, and to be mirror-symmetric[7] w.r.t. the xz plane. The beam is subjected to a line load $\bar{q}(x) = q(x)\bar{e}_z$ in its cross-direction (Fig. 5.7).

[7] *Mirror-symmetric* – composed of two parts, being reflections of each other.

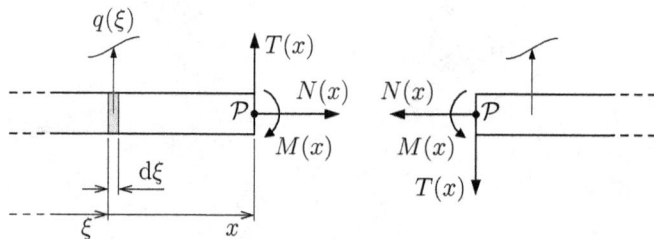

For this planar geometry, we place a section Π_x trough a point $\mathcal{P}$ at a position x along the beam. Then, the section force and the section couple acting in $\mathcal{P}$ become

$$\bar{F}_\mathcal{P}(x) = N(x)\bar{e}_x + T(x)\bar{e}_z,$$
$$C_\mathcal{P}(x) = M(x)\bar{e}_y,$$

where the components are called the *normal force $N(x)$*, the *shear force $T(x)$*, and the *bending moment $M(x)$*. On the section surface with normal $\bar{e}_x$, these internal forces and internal couple are defined as positive in the coordinate directions. Conversely, it follows from the Law of action and reaction that the internal forces and internal couple are negative in the coordinate directions on the opposite section surface with normal $-\bar{e}_x$ (Fig. 5.8).

Figure 5.8: The forces and couple acting in a section. The section is located at x, while the line load is parameterized by a coordinate ξ along the x axis.

Knowing the internal forces and couple, $N(x)$, $T(x)$ and $M(x)$, is of crucial importance when designing beam structures according to strength requirements.

Static equilibrium of planar beams

If a beam is mounted on supports and loaded by a given line load $q(x)$, then the shear force and the bending moment along the beam can be determined using the following method:

1. Place a section at an arbitrary position x along the beam.

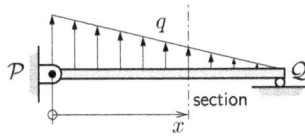

2. Draw a free-body diagram for the entire beam. Static equilibrium equations give the forces and couples at the supports.

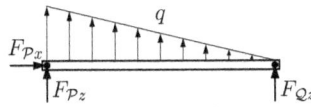

3. Draw a free-body diagram for one part of the beam. Static equilibrium equations give the internal forces and couple.

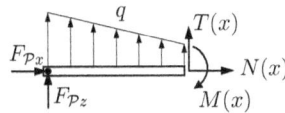

An alternative method for determining the shear force and the bending moment is based on differential equations relating $T(x)$ and $M(x)$ to $q(x)$.

Theorem 5.4 (Equilibrium for beams)**.** For a beam aligned with the x direction, and loaded with a line load $\bar{q}(x) = q(x)\bar{e}_z$, it holds that

$$\frac{\mathrm{d}T}{\mathrm{d}x} = -q(x), \tag{5.6a}$$

$$\frac{\mathrm{d}M}{\mathrm{d}x} = T(x), \tag{5.6b}$$

where $T(x)$ is the shear force, and $M(x)$ is the bending moment of the cross-section of the beam.

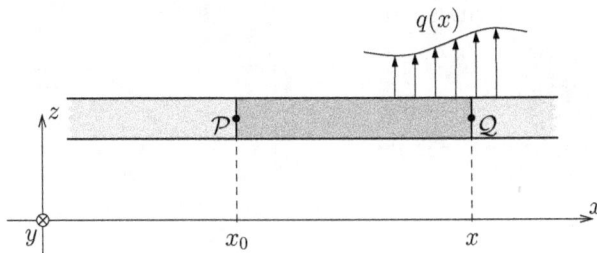

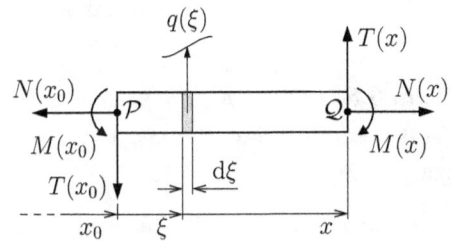

Figure 5.9: A free-body diagram of a beam segment $\mathcal{PQ}$ from x_0 to x.

Proof. Consider the beam segment $\mathcal{PQ}$ from x_0 to x, with its free-body diagram in Fig. 5.9. A force equilibrium for $\mathcal{PQ}$ in the z direction gives

$$T(x) - T(x_0) + \int_{x_0}^{x} q(\xi)\mathrm{d}\xi = 0 \quad \Leftrightarrow \quad \left\{\frac{\mathrm{d}}{\mathrm{d}x}, \text{ Eq. (A.37)}\right\} \quad \Leftrightarrow$$

$$\frac{\mathrm{d}T}{\mathrm{d}x} + q(x) = 0. \tag{5.7}$$

A counterclockwise moment equilibrium for $\mathcal{PQ}$, w.r.t. $\mathcal{P}$, gives

$$(x - x_0)T(x) - M(x) + M(x_0) + \int_{x_0}^{x} (\xi - x_0)q(\xi)\mathrm{d}\xi = 0 \quad \Leftrightarrow \quad \left\{ \frac{\mathrm{d}}{\mathrm{d}x}, \text{Eq. (A.37)} \right\} \quad \Leftrightarrow$$

$$T(x) + (x - x_0)\frac{\mathrm{d}T}{\mathrm{d}x} - \frac{\mathrm{d}M}{\mathrm{d}x} + (x - x_0)q(x) = 0, \quad \Leftrightarrow \quad \{\text{Eq. (5.7)}\} \quad \Leftrightarrow$$

$$T(x) - \frac{\mathrm{d}M}{\mathrm{d}x} = 0. \qquad \square$$

One additional differential equation is obtained by differentiation of Eq. (5.6b) w.r.t. x, and subsequent insertion of Eq. (5.6a):

$$\frac{\mathrm{d}^2 M}{\mathrm{d}x^2} = -q(x). \tag{5.8}$$

Boundary conditions are required to solve the differential equations (5.6a), (5.6b) or (5.8). If a beam $\mathcal{PQ}$ in the interval $x \in [x_0, x_1]$ is loaded by forces and couples at its endpoints, then the boundary conditions become (Fig. 5.10)

$$\begin{cases} T(x_1) = +F_{\mathcal{Q}z}, \\ M(x_1) = +C_{\mathcal{Q}y}, \end{cases} \qquad \begin{cases} T(x_0) = -F_{\mathcal{P}z}, \\ M(x_0) = -C_{\mathcal{P}y}. \end{cases}$$

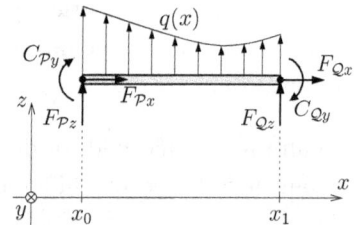

Figure 5.10: Geometry for the boundary conditions of the differential equations for the static equilibrium of beams.

Thus, the boundary values of the shear force equal the cross-directional components of the external forces at the endpoints, while obeying the specified sign convention (Fig. 5.8). Similarly, the bending moment equals the external couple.

5.3 Fluid statics

Liquids and gases lack the ability to preserve their shape. Thus, according to Def. 1.1, a body that consists of a liquid or a gas cannot be a rigid body. A *fluid* is an idealized material suitable for modeling liquids and gases.

Definition 5.5 (Fluid). A *fluid* is a continuous material such that, when in static equilibrium, the tangential surface load on every section through the fluid is zero.

Fluid statics concerns fluids in static equilibrium (Def. 3.1). The internal forces of a fluid are described by a scalar-valued function called *pressure*.

Definition 5.6 (Pressure). For a fluid in static equilibrium, the surface load on a planar section Π_λ is $-p(\bar{r}, \bar{e}_\lambda)\bar{e}_\lambda$, with $\bar{e}_\lambda$ the normal direction of Π_λ, and $\bar{r}$ a position vector (Fig. 5.11). The *pressure* is the scalar-valued function $p(\bar{r}, \bar{e}_\lambda)$.

Pressure has the SI unit of pascal (Pa), or derived USC units:

$$1\,\text{Pa} = 1\,\frac{\text{N}}{\text{m}^2} = 1\,\frac{\text{kg}}{\text{m·s}^2}, \qquad 1\,\frac{\text{lb}_\text{f}}{\text{ft}^2} = \frac{1}{144}\,\frac{\text{lb}_\text{f}}{\text{in}^2} = \frac{1}{144}\,\text{psi} \approx 47.880\,\text{Pa},$$

where the unit pound-force per square inch (psi) is commonly used with USC.

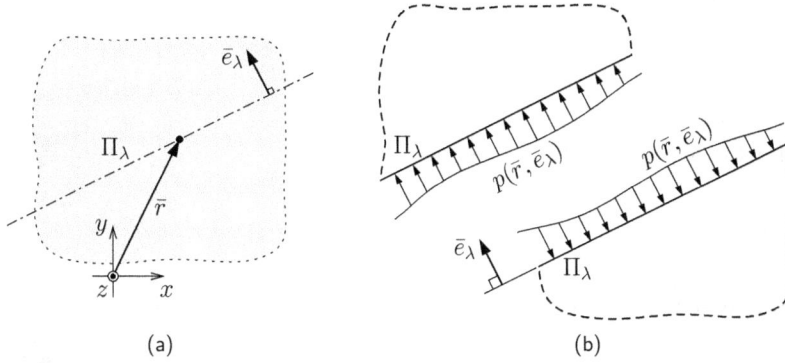

(a) (b)

Figure 5.11: (a) A planar section Π_λ, through a fluid in static equilibrium. (b) The section Π_λ is subjected to a surface load $-p(\bar{r}, \bar{e}_\lambda)\bar{e}_\lambda$, which is oriented in the inward normal direction of the section surface.

According to the Law of action and reaction, the pressure is the same, but oppositely directed, for opposite section surfaces (Fig. 5.11b). In Def. 5.6, the pressure may depend on the orientation $\bar{e}_\lambda$ of the section. It turns out, however, that the pressure is independent of $\bar{e}_\lambda$. To prove this, we first formulate a *lemma*.

Lemma 5.7 (Projected area of a parallelogram). The projected area of a parallelogram, with unit normal $\bar{e}_\lambda$ and area $A_\lambda > 0$, onto a plane with unit normal $\bar{e}$ is

$$A = A_\lambda |\bar{e}_\lambda \cdot \bar{e}| = A_\lambda |\cos\varphi|, \tag{5.9}$$

where φ is the angle between the normals of the parallelogram and the plane (Fig. 5.12).

Proof. Let $\bar{u}$ and $\bar{v}$ be geometric vectors which define the sides of the parallelogram, so that

$$\bar{u} \times \bar{v} = A_\lambda \bar{e}_\lambda. \tag{5.10}$$

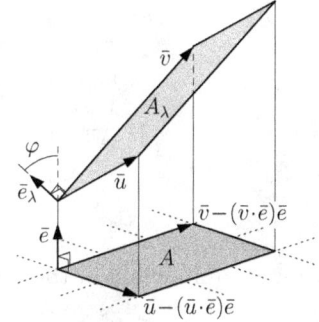

Figure 5.12: Geometry for Lemma 5.7.

The projections of the vectors onto the plane with normal $\bar{e}$ are $\bar{u} - (\bar{u} \cdot \bar{e})\bar{e}$ and $\bar{v} - (\bar{v} \cdot \bar{e})\bar{e}$, respectively, so that the projected area of the parallelogram is

$$
\begin{aligned}
A &= \left| [\bar{u} - (\bar{u} \cdot \bar{e})\bar{e}] \times [\bar{v} - (\bar{v} \cdot \bar{e})\bar{e}] \right| \\
&= \left| \bar{u} \times \bar{v} - (\bar{u} \cdot \bar{e})\bar{e} \times \bar{v} - \bar{u} \times (\bar{v} \cdot \bar{e})\bar{e} + (\bar{u} \cdot \bar{e})(\bar{v} \cdot \bar{e})\bar{e} \times \bar{e} \right| = \left\{ \bar{e} \times \bar{e} = \bar{0}, \text{ Eq. (A.21a)} \right\} \\
&= \left| \bar{u} \times \bar{v} + (\bar{u} \cdot \bar{e})\bar{v} \times \bar{e} - (\bar{v} \cdot \bar{e})\bar{u} \times \bar{e} \right| \\
&= \left| \bar{u} \times \bar{v} - [(\bar{e} \cdot \bar{v})\bar{u} - (\bar{e} \cdot \bar{u})\bar{v}] \times \bar{e} \right| = \left\{ \text{Eq. (A.22a)} \right\} \\
&= \left| \bar{u} \times \bar{v} - [\bar{e} \times (\bar{u} \times \bar{v})] \times \bar{e} \right| = \left\{ \text{Eq. (A.22a)} \right\} \\
&= \left| \bar{u} \times \bar{v} - \{ (\bar{e} \cdot \bar{e})(\bar{u} \times \bar{v}) - [\bar{e} \cdot (\bar{u} \times \bar{v})]\bar{e} \} \right| = \left\{ \bar{e} \cdot \bar{e} = 1 \right\} \\
&= \left| (\bar{u} \times \bar{v}) \cdot \bar{e} \right| |\bar{e}| = \left\{ \text{Eq. (5.10)} \right\} \\
&= A_\lambda |\bar{e}_\lambda \cdot \bar{e}| \qquad\qquad\qquad \square
\end{aligned}
$$

Theorem 5.8 (Pascal's principle). For a fluid in static equilibrium in a constant field of gravity $\bar{g}$, it holds that the pressure $p(\bar{r}, \bar{e}_\lambda) = p(\bar{r})$ on a section Π_λ is independent of the orientation $\bar{e}_\lambda$ of the section surface.

Proof. Choose an arbitrary point $\mathcal{P}$ and an arbitrary direction $\bar{e}_\lambda$. Moreover, choose a coordinate system xyz such that $\bar{g} = -g\bar{e}_y$ and $\mathcal{P}$ is located on the positive z axis. Let Ω be an arbitrary open domain in the xy plane. Consider the body

$$\{(x, y, z) : (x, y) \in \Omega, \ 0 \le z \le z_\lambda(x, y)\},$$

where $z = z_\lambda(x, y)$ is the equation for the planar section Π_λ that intersects $\mathcal{P}$ (Fig. 5.13).

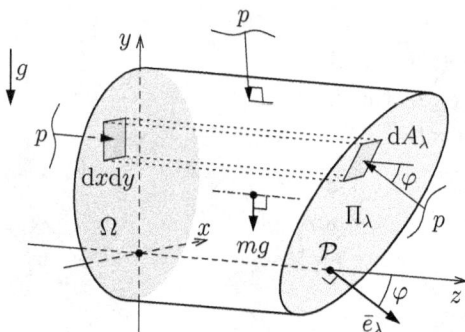

Figure 5.13: Geometry for Theorem 5.8. Since the force of gravity and the pressure on the curved surface are perpendicular to $\bar{e}_z$, they do not enter into the force equilibrium in the z direction.

A force equilibrium in the z direction for this body gives

$$\int_\Omega p(x, y, 0, \bar{e}_z)\mathrm{d}x\mathrm{d}y - \int_\Omega p[x, y, z_\lambda(x, y), \bar{e}_\lambda] \cos\varphi \mathrm{d}A_\lambda = 0, \quad \forall\Omega, \quad \Leftrightarrow \quad \left\{ \begin{array}{l} \text{Lemma 5.7,} \\ \cos\varphi\mathrm{d}A_\lambda = \mathrm{d}x\mathrm{d}y \end{array} \right\} \quad \Leftrightarrow$$

$$\int_\Omega \{p(x, y, 0, \bar{e}_z) - p[x, y, z_\lambda(x, y), \bar{e}_\lambda]\} \, \mathrm{d}x\mathrm{d}y = 0, \quad \forall\Omega, \quad \Leftrightarrow \quad \{\text{Localization, Theorem A.6}\} \quad \Leftrightarrow$$

$$p(x, y, 0, \bar{e}_z) - p[x, y, z_\lambda(x, y), \bar{e}_\lambda] = 0.$$

We choose $x = y = 0$ and use that $z_\lambda(0, 0) = z_\mathcal{P}$, giving

$$p(0, 0, z_\mathcal{P}, \bar{e}_\lambda) = p(0, 0, 0, \bar{e}_z).$$

Hence, the pressure in the arbitrary point $\mathcal{P}$ is independent of the orientation $\bar{e}_\lambda$ of the section. $\qquad\square$

According to Pascal's principle 5.8, the pressure only depends on the position. *Pascal's law* specifies how this pressure varies with position in a constant field of gravity.

Theorem 5.9 (Pascal's law). For a fluid with constant density ϱ, in static equilibrium in a constant field of gravity $\bar{g}$, it holds that

$$p_\mathcal{Q} = p_\mathcal{P} + \varrho\bar{g} \cdot \overline{\mathcal{P}\mathcal{Q}}, \tag{5.11}$$

where $\mathcal{P}$ and $\mathcal{Q}$ are material points, and where $p_{\mathcal{P}}$ and $p_{\mathcal{Q}}$ are the respective pressures in those points.

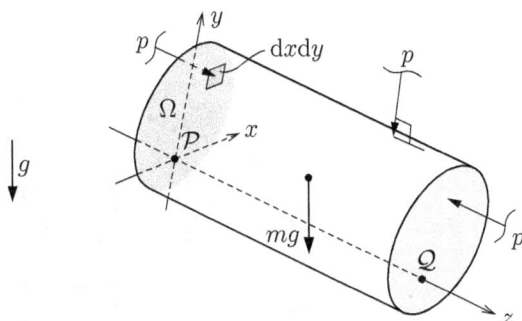

Figure 5.14: Geometry for Theorem 5.9. The cross-section of the cylinder does not have to be circular.

Proof. We introduce a rectangular coordinate system xyz with origin in $\mathcal{P}$, such that $\mathcal{Q}$ has the coordinates $(0,0,\ell)$. Let Ω be an arbitrary open domain in the xy plane, and consider the perpendicular cylinder

$$\{(x,y,z) : (x,y) \in \Omega,\ 0 \le z \le \ell\},$$

whose mass is denoted by m (Fig. 5.14). A force equilibrium for this cylinder in the z direction gives

$$\int_{\Omega} p(x,y,0)\mathrm{d}x\mathrm{d}y - \int_{\Omega} p(x,y,\ell)\mathrm{d}x\mathrm{d}y + m\bar{g}\cdot\bar{e}_z = 0,\ \ \forall\Omega,\ \ \Leftrightarrow\ \ \left\{ m = \varrho\ell \int_{\Omega} \mathrm{d}x\mathrm{d}y \right\}\ \ \Leftrightarrow$$

$$\int_{\Omega} [p(x,y,0) - p(x,y,\ell)]\,\mathrm{d}x\mathrm{d}y + \varrho\bar{g}\cdot\ell\bar{e}_z \int_{\Omega} \mathrm{d}x\mathrm{d}y = 0,\ \ \forall\Omega,\ \ \Leftrightarrow\ \ \{\ell\bar{e}_z = \overline{\mathcal{PQ}}\}\ \ \Leftrightarrow$$

$$\int_{\Omega} \left[p(x,y,0) - p(x,y,\ell) + \varrho\bar{g}\cdot\overline{\mathcal{PQ}} \right]\mathrm{d}x\mathrm{d}y = 0,\ \ \forall\Omega,\ \ \Leftrightarrow\ \ \{\text{Localization, Theorem A.6}\}\ \ \Leftrightarrow$$

$$p(x,y,0) - p(x,y,\ell) + \varrho\bar{g}\cdot\overline{\mathcal{PQ}} = 0.$$

Particularly, by choosing $x = y = 0$, we obtain

$$p(0,0,0) - p(0,0,\ell) + \varrho\bar{g}\cdot\overline{\mathcal{PQ}} = 0\ \ \Leftrightarrow$$

$$p_{\mathcal{P}} - p_{\mathcal{Q}} + \varrho\bar{g}\cdot\overline{\mathcal{PQ}} = 0. \qquad \square$$

Pascal's law 5.9 was shown for a situation when the line $\mathcal{PQ}$ is embedded in fluid. If $\mathcal{P}$ and $\mathcal{Q}$ have an obstruction between them, while the fluid still occupies a connected region, then a polyline $\mathcal{P}\mathcal{A}_1\mathcal{A}_2\cdots\mathcal{A}_n\mathcal{Q}$ can be constructed in the interior of the fluid (Fig. 5.15), so that

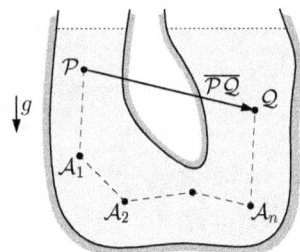

$$p_{\mathcal{A}_1} = p_{\mathcal{P}} + \varrho\bar{g}\cdot\overline{\mathcal{P}\mathcal{A}_1}$$

$$p_{\mathcal{A}_2} = p_{\mathcal{A}_1} + \varrho\bar{g}\cdot\overline{\mathcal{A}_1\mathcal{A}_2}$$

$$\vdots$$

$$p_{\mathcal{Q}} = p_{\mathcal{A}_n} + \varrho\bar{g}\cdot\overline{\mathcal{A}_n\mathcal{Q}}.$$

Figure 5.15: A polyline can be constructed between $\mathcal{P}$ and $\mathcal{Q}$ in the interior of a fluid.

A summation of these equations recovers Pascal's law:

$$p_{\mathcal{Q}} + \sum_{i=1}^{n} p_{\mathcal{A}_i} = p_{\mathcal{P}} + \sum_{i=1}^{n} p_{\mathcal{A}_i} + \varrho \bar{g} \cdot (\overline{\mathcal{P}\mathcal{A}_1} + \overline{\mathcal{A}_1\mathcal{A}_2} + \cdots + \overline{\mathcal{A}_n\mathcal{Q}}) \quad \Leftrightarrow$$

$$p_{\mathcal{Q}} = p_{\mathcal{P}} + \varrho \bar{g} \cdot \overline{\mathcal{P}\mathcal{Q}}. \qquad \square$$

Since Pascal's law requires a constant fluid density, it is typically applicable for liquids. If a liquid is in contact with atmospheric pressure p_0, a *depth coordinate* h can be introduced in the direction of the field of gravity, with its origin at the free surface of the liquid (Fig. 5.16). Thus, Pascal's law 5.9 can be written

$$p(h) = p_0 + \varrho g h. \qquad (5.12)$$

One consequence of Eq. (5.12) is that, in a connected liquid in static equilibrium, every free surface at atmospheric pressure has the same depth coordinate $h = 0$ (Fig. 5.16).

The pressure on the surface of a region in the interior of a fluid, can be represented by a buoyancy force that acts on this region. This upward force balances the force of gravity that acts within the same region:

Theorem 5.10. The pressure on the surface of a region Ω of fluid in static equilibrium can be represented by a force

$$\bar{F}_{\mathcal{C}} = -\varrho V \bar{g}, \qquad (5.13)$$

that acts on Ω in its centroid $\mathcal{C}$. Here, $\bar{g}$ is a constant field of gravity, ϱ is the constant density of the fluid, and V is the volume of Ω.

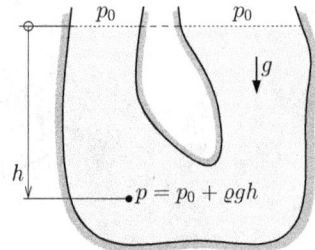

Figure 5.16: The depth coordinate h in the direction of the field of gravity, with its origin at the surface of the liquid. For a connected domain of liquid, all free liquid surfaces at depth $h = 0$ have the same pressure p_0.

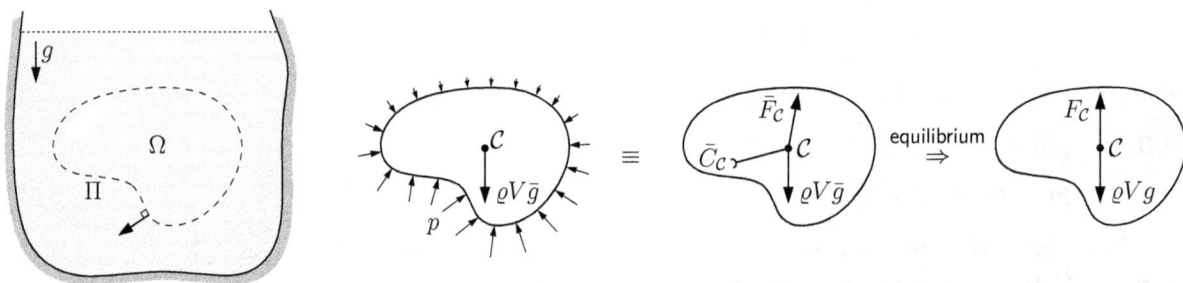

Figure 5.17: A free-body diagram of a region Ω is drawn and the pressure on Π is reduced w.r.t. the centroid $\mathcal{C}$ of Ω.

Proof. If we place a section $\Pi = \partial\Omega$ with an outward normal on the boundary of Ω, then the pressure on Π can be reduced to a force $\bar{F}_{\mathcal{C}}$ and a couple $\bar{C}_{\mathcal{C}}$, which act in the centroid $\mathcal{C}$ of Ω (Fig. 5.17). The only additional force acting on Ω is the force of gravity $\varrho V \bar{g}$ with point of application $\mathcal{C}$. A moment equilibrium for Ω w.r.t. $\mathcal{C}$ gives

$$\bar{C}_{\mathcal{C}} = \bar{0},$$

and a force equilibrium gives

$$\bar{F}_C + \varrho V \bar{g} = \bar{0}.$$ □

Theorem 5.11 (Archimedes' principle). The pressure on the surface of a body Ω, submerged in a fluid in static equilibrium, can be represented by a buoyancy force

$$\bar{F}_C = -\varrho V \bar{g}, \tag{5.14}$$

that acts on this body in its centroid C. Here, $\bar{g}$ is a constant field of gravity, ϱ is the constant density of the fluid, and V is the volume of Ω.

Proof. From Pascal's law 5.9 it follows that the pressure in every point on the boundary $\partial\Omega$ of the body is independent of the type of material inside Ω. Therefore, the material inside Ω can be replaced by fluid without changing the pressure on $\partial\Omega$. For this equivalent physical system, Theorem 5.10 shows that the pressure on $\partial\Omega$ can be represented by a force $\bar{F}_C = -\varrho V \bar{g}$, acting in the centroid C of Ω.   □

When a body with mass m is submerged in liquid, the force of gravity $m\bar{g}$ acts in its center of mass $\mathcal{G}$. At the same time, due to the pressure of the fluid, a buoyancy force $-\varrho V \bar{g}$ acts in the centroid C of the body. Note that ϱV is the mass of the displaced fluid. As usual, additional contact forces may act on the body (Fig. 5.18).

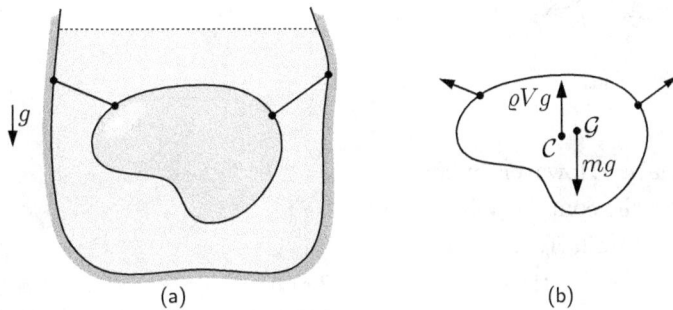

(a)   (b)

Figure 5.18: (a) A body with nonuniformly distributed mass is submerged in a fluid. (b) A free-body diagram of this body.

6

Friction

Friction forces arise at contact places between bodies. This friction counteracts sliding of contacting surfaces relative to each other.[8] Consider two bodies, Ω_1 and Ω_2, that are in physical contact at a common point $\mathcal{P}$ (Fig. 6.1). A *tangent plane* of the bodies is defined at this contact point $\mathcal{P}$, having a normal vector $\bar{e}_n$. A normal force $\bar{N} = N\bar{e}_n$ and a friction force $\bar{F}_f \perp \bar{e}_n$ act on Ω_1. Conversely, according to the Law of action and reaction, a normal force $-\bar{N}$ and a friction force $-\bar{F}_f$ act on Ω_2.

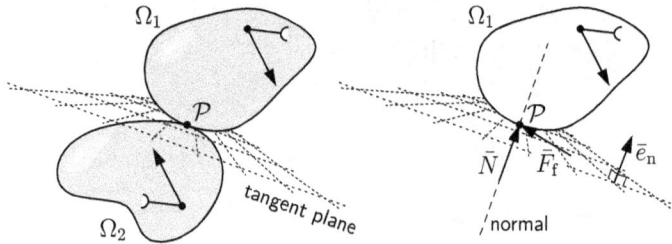

Friction arises between all types of materials. However, when the friction between two bodies is deemed negligible, the contact point is said to be *frictionless* with a friction force $\bar{F}_f = \bar{0}$. Similarly, a *frictionless surface*[9] is a surface for which all contact points are frictionless.

6.1 A friction experiment

Consider the experimental setup in Fig. 6.2. A solid box rests on a carriage placed on a horizontal surface. A varying horizontal force P, whose magnitude is measured using a force sensor, acts on this box. The carriage is fixed in place by a device that measures the magnitude F of the horizontal force on the carriage. By formulating force equilibrium for the carriage, we obtain the friction force $F_f = F$, as shown by the free-body diagrams in Fig. 6.2.

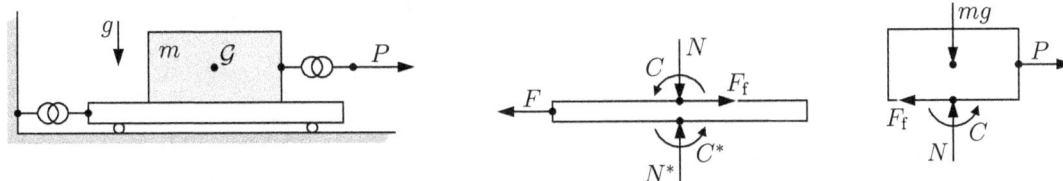

Figure 6.2: An illustration of a friction experiment. Two force sensors have been indicated by double-circle symbols. Free-body diagrams show the carriage and the box.

In an experiment, a force $P = 0$ initially acts on a box, and then P is slowly increased. As long as P is sufficiently small, the box does not slide against the horizontal surface of the carriage. It is held in position by *static friction*. While sliding does not occur, a force equilibrium is maintained, giving $F_f = P$. However, when P reaches a limiting value, the box starts to slide against the carriage, and to accelerate. At this transition, the friction force decreases somewhat to a constant value that is essentially unaffected by changes in P while the box moves (Fig. 6.3). Friction during sliding is called *kinetic friction*.

The behavior observed in the above experiment is typical to *dry friction*, *i.e.* when contact surfaces are dry and clean. Moisture, micro-particles or oxide layers on the surfaces of the bodies affect friction. Likewise, temperature or bulk mechanical properties of the bodies can affect friction.

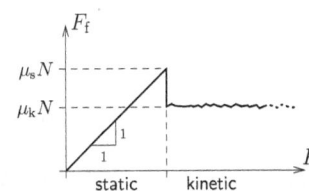

Figure 6.3: The friction force plotted as a function of the applied force P for the experiment in Fig. 6.2.

6.2 Coulomb friction

If we restrict our discussion to dry friction between clean surfaces, the following empirical relation[10] approximately holds:

[10] *Empirical relation* – an equation that is motivated only by experimental evidence.

Empirical relation 6.1 (Coulomb friction). If a place of contact is in a state of static friction, then static friction persists as long as

$$\frac{|\bar{F}_f|}{N} < \mu_s. \tag{6.1}$$

When sliding occurs at a place of contact, we have

$$|\bar{F}_f| = \mu_k N. \tag{6.2}$$

Here, $\bar{F}_f$ is the friction force, N is the magnitude of the normal force, μ_s is the coefficient of static friction, and μ_k is the coefficient of kinetic friction, where $0 \leq \mu_k \leq \mu_s$.

During sliding, $\bar{F}_f$ counteracts the sliding motion. The thought experiment in Sect. 6.1 (Figs. 6.2 and 6.3) is a prototypical example of Coulomb friction.

If a place of contact is in a state of static friction, it is often of interest to investigate the limit of *sliding initiation*. This is achieved by setting the friction force to the limiting value at which a sliding transition is to occur:

$$|\bar{F}_f| = \mu_s N. \tag{6.3}$$

This corresponds to the maximum of the friction force in Fig. 6.3.

When solving problems that include Coulomb friction, it is not always known whether sliding occurs at a place of contact. In such cases, one first assumes a state of static friction, and uses the equilibrium Eqs. (3.1a) and (3.1b) to determine the friction force $\bar{F}_f$ and the magnitude of the normal force N, conditional on static friction. If this result violates Eq. (6.1), then sliding must occur, implying that the friction force is instead given by Eq. (6.2), and that static equilibrium no longer holds.

6.3 Friction in multi-body systems

If there are several places of contact governed by Coulomb friction in a multi-body system, then the empirical relation 6.1 holds for each place of contact. For instance, if we have two places of contact, $\mathcal{P}$ and $\mathcal{Q}$, then the following outcomes are conceivable (Fig. 6.4):

- There is no sliding at $\mathcal{P}$, nor at $\mathcal{Q}$.
- There is sliding at $\mathcal{P}$, but not at $\mathcal{Q}$.
- There is sliding at $\mathcal{Q}$, but not at $\mathcal{P}$.
- There is sliding at both $\mathcal{P}$ and $\mathcal{Q}$.

If a multi-body system includes n places of contact, then there are 2^n possible combinations of sliding and static friction.

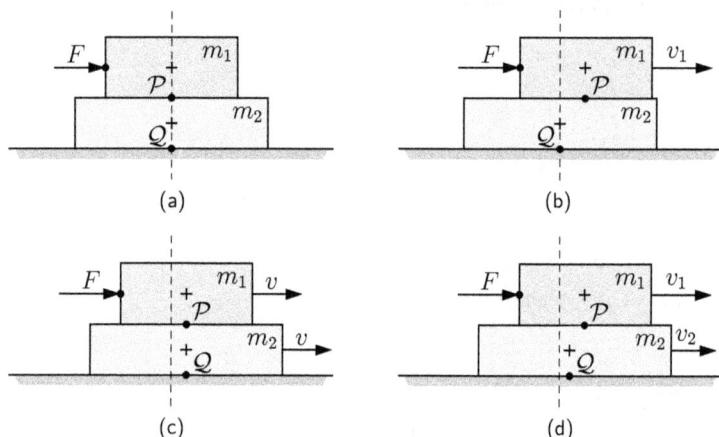

Figure 6.4: An example of friction at several places of contact. The possible outcomes are: (a) No sliding occurs, (b) sliding only occurs at $\mathcal{P}$, (c) sliding only occurs at $\mathcal{Q}$, and (d) sliding occurs at both $\mathcal{P}$ and $\mathcal{Q}$.

Sometimes, the geometric constraints of a mechanical system permit for some combinations of sliding and static friction to be eliminated. For

Figure 6.5: To push a wedge into a log, simultaneous sliding is required at both its places of contact.

instance, a wedge has two contact surfaces (Fig. 6.5). The wedge can only move if simultaneous sliding occurs at the contact surfaces. For this reason, there are only two possible scenarios: either sliding occurs at both places of contact, or static friction is maintained at both places of contact.

6.4 Belt friction

When a belt is wound around a cylinder surface, friction counteracts sliding between the belt and the cylinder. As a consequence, the tensile force T can be different in the two ends of the belt (Fig. 6.6).

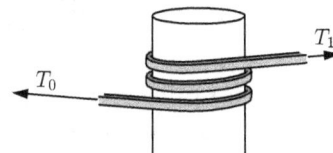

Figure 6.6: When a belt is wound around the surface of a cylinder, friction allows for a difference between the tensile force, at the ends of the belt.

Theorem 6.2 (Belt friction). For a massless belt sliding against a cylindrical surface with the coefficient μ_k of kinetic friction, the tensile force in the belt is

$$T(\theta) = T_0 e^{-\mu_k \theta}, \tag{6.4}$$

where $\theta \geq 0$ is the polar angle, $T_0 = T(0)$ is a constant, and the sliding direction is opposite to the polar angle direction (Fig. 6.7).

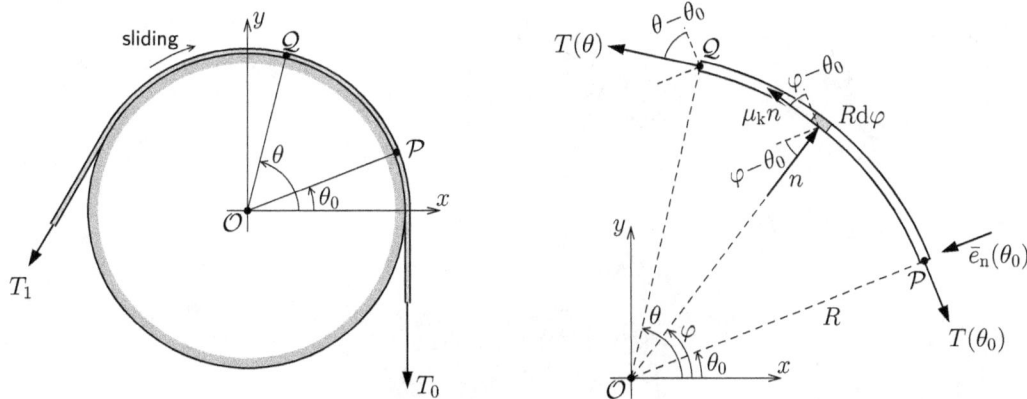

Figure 6.7: A belt slides along the surface of a cylinder. A free-body diagram is drawn for the segment $\mathcal{PQ}$ of the belt, showing the friction force $\mu_k n$ whose direction is opposite to the sliding direction.

Proof. Let R denote the radius of the cylinder surface. The normal force acting on the belt is a line force $n(\theta)$. From Eq. (6.2), the friction force is $\mu_k n(\theta)$ against the sliding direction. Consider a segment $\mathcal{PQ}$ of the belt ranging from θ_0 to θ, as drawn in the free-body diagram of Fig. 6.7. A force equilibrium for $\mathcal{PQ}$ in the normal direction $\bar{e}_n(\theta_0)$ gives

$$T(\theta)\sin(\theta-\theta_0) + \int_{\theta_0}^{\theta} \left[\mu_k n(\varphi)\sin(\varphi-\theta_0) - n(\varphi)\cos(\varphi-\theta_0)\right] R\mathrm{d}\varphi = 0, \quad \forall \theta_0, \theta \quad \Leftrightarrow \quad \left\{\frac{\mathrm{d}}{\mathrm{d}\theta}, \text{Eq. (A.37)}\right\} \quad \Leftrightarrow$$

$$\frac{\mathrm{d}T}{\mathrm{d}\theta}\sin(\theta-\theta_0) + T(\theta)\cos(\theta-\theta_0) + \mu_k R n(\theta)\sin(\theta-\theta_0) - R n(\theta)\cos(\theta-\theta_0) = 0, \quad \forall \theta_0, \theta. \tag{6.5}$$

Equation (6.5) holds for all θ_0 and θ, including $\theta_0 \to \theta$, which gives:

$$T(\theta) - Rn(\theta) = 0. \qquad (6.6)$$

By inserting Eq. (6.6) into (6.5), we obtain

$$\frac{\mathrm{d}T}{\mathrm{d}\theta} \sin(\theta - \theta_0) + \mu_\mathrm{k} T(\theta) \sin(\theta - \theta_0) = 0, \quad \forall \theta_0, \theta \quad \Leftrightarrow$$

$$\frac{\mathrm{d}T}{\mathrm{d}\theta} + \mu_\mathrm{k} T(\theta) = 0. \qquad (6.7)$$

The solution to this differential equation (6.7), with the boundary condition $T(0) = T_0$, is $T(\theta) = T_0 e^{-\mu_\mathrm{k}\theta}$. $\qquad \square$

The relation between the tensile forces T_0 and T_1, at the respective ends of the belt, is given by inserting the total angle of application β of the belt into Eq. (6.4):

$$T_1 = T(\beta) = T_0 e^{-\mu_\mathrm{k}\beta}. \qquad (6.8)$$

With an argument similar to that of Theorem 6.2, it can be shown that the condition for static friction for the belt is

$$T_0 e^{-\mu_\mathrm{s}\beta} < T_1 < T_0 e^{\mu_\mathrm{s}\beta}, \qquad (6.9)$$

where consideration has been given to the two possible sliding directions.

PART II
PARTICLE DYNAMICS

7
Planar kinematics of particles

Kinematics concerns the geometry of motion, without considering the forces that cause motion. This chapter focuses on the motion of particles or points confined to a plane, known as *planar motion*.

7.1 Rectilinear motion

A particle $\mathcal{P}$ moving along a straight line in space, is said to describe *rectilinear motion*. To represent the position of $\mathcal{P}$, we introduce a position coordinate $x(t)$, relative to a fixed point $\mathcal{O}$ on the line of motion (Fig. 7.1). Consequently, $x(t)$ represents a unique position for each time t, and $x(t)$ may assume both positive and negative values. The instantaneous *velocity* of the particle is defined from its mean velocity between times t and $t + \Delta t$ by taking the limit

$$v(t) \equiv \lim_{\Delta t \to 0} \frac{x(t + \Delta t) - x(t)}{\Delta t} = \frac{dx}{dt}, \tag{7.1}$$

which is the time derivative of the position $x(t)$. For rectilinear motion, the *speed* of the particle is defined as $|v|$.

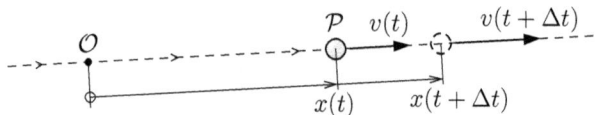

Figure 7.1: The motion of a particle $\mathcal{P}$ along a straight line, relative to a fixed reference point $\mathcal{O}$.

Similarly, the instantaneous *acceleration* of a particle is defined as the time derivative of the velocity:

$$a(t) \equiv \lim_{\Delta t \to 0} \frac{v(t + \Delta t) - v(t)}{\Delta t} = \frac{dv}{dt}. \tag{7.2}$$

Alternatively, the definitions of velocity and acceleration can be written using differentials (Appendix A.3). By applying Eq. (A.30) to Eqs.

(7.1) and (7.2), respectively, we obtain

$$dx = vdt, \qquad\qquad (7.3a)$$

$$dv = adt. \qquad\qquad (7.3b)$$

Theorem 7.1. For the rectilinear motion of a particle, with position coordinate $x(t)$, velocity $v(t)$, and acceleration $a(t)$, it holds that

$$vdv = adx. \qquad\qquad (7.4)$$

Proof. From Eq. (7.3b) we obtain

$$dv = adt \quad \Leftrightarrow \quad \{\text{multiply by } v\} \quad \Leftrightarrow$$

$$vdv = avdt \quad \Leftrightarrow \quad \{\text{Eq. (7.3a)}\} \quad \Leftrightarrow$$

$$vdv = adx. \qquad\qquad \square$$

To solve kinematics problems, one typically proceeds from one or several of the differential relations (7.3a), (7.3b) and (7.4). Scalar equations are then formed by integrating the differential relations according to Theorems A.2 or A.3.

7.2 Curvilinear motion

For general particle motion, the time-dependent position of a particle, or a point in space, is denoted by $\bar{r}(t)$. From this position vector, the velocity and the acceleration are again defined as limits.

Definition 7.2 (Velocity). The velocity of a particle with position vector $\bar{r}(t)$ is (Fig. 7.2)

$$\bar{v}(t) \equiv \lim_{\Delta t \to 0} \frac{\bar{r}(t + \Delta t) - \bar{r}(t)}{\Delta t} = \lim_{\Delta t \to 0} \frac{\Delta \bar{r}}{\Delta t} = \frac{d\bar{r}}{dt}. \qquad (7.5)$$

The velocity is a vector quantity, whose direction is tangent to the path described by $\bar{r}(t)$ (Fig. 7.2).

Definition 7.3 (Acceleration). The acceleration of a particle, with velocity $\bar{v}(t)$, is (Fig. 7.3)

$$\bar{a}(t) \equiv \lim_{\Delta t \to 0} \frac{\bar{v}(t + \Delta t) - \bar{v}(t)}{\Delta t} = \lim_{\Delta t \to 0} \frac{\Delta \bar{v}}{\Delta t} = \frac{d\bar{v}}{dt}. \qquad (7.6)$$

The acceleration is a vector quantity, whose direction is not necessarily tangent to the path described by $\bar{r}(t)$.

Rectangular coordinates

In a rectangular coordinate system xyz with basis $\{\bar{e}_x, \bar{e}_y, \bar{e}_z\}$, the position vector of a particle is written

$$\bar{r}(t) = x(t)\bar{e}_x + y(t)\bar{e}_y + z(t)\bar{e}_z. \qquad (7.7)$$

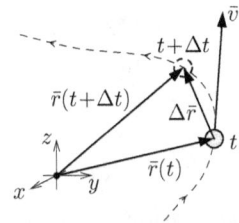

Figure 7.2: Geometry for the definition of velocity as a limit.

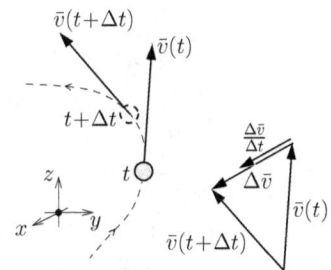

Figure 7.3: Geometry for the definition of acceleration as a limit.

This situation is illustrated in Fig. 7.4. If it is clear from the context that a quantity is time-dependent, then one often omits the parameter t and simply writes $\bar{r} = x\bar{e}_x + y\bar{e}_y + z\bar{e}_z$.

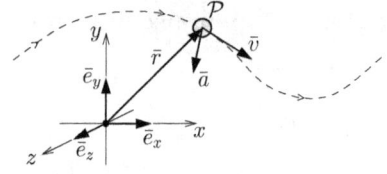

Figure 7.4: The motion of a particle $\mathcal{P}$ in space, relative to a rectangular coordinate system.

Theorem 7.4 (Velocity in rectangular coordinates). In a rectangular coordinate system, the velocity of a particle with position vector $\bar{r} = x\bar{e}_x + y\bar{e}_y + z\bar{e}_z$ is

$$\bar{v} = \dot{x}\bar{e}_x + \dot{y}\bar{e}_y + \dot{z}\bar{e}_z. \tag{7.8}$$

Proof. According to Def. 7.2 concerning the velocity, we have

$$\begin{aligned}
\bar{v} &= \frac{d\bar{r}}{dt} \\
&= \frac{d}{dt}\left(x\bar{e}_x + y\bar{e}_y + z\bar{e}_z\right) = \left\{\text{Product rule (A.25a)}\right\} \\
&= \dot{x}\bar{e}_x + x\frac{d\bar{e}_x}{dt} + \dot{y}\bar{e}_y + y\frac{d\bar{e}_y}{dt} + \dot{z}\bar{e}_z + z\frac{d\bar{e}_z}{dt} = \left\{\bar{e}_x, \bar{e}_y, \bar{e}_z \text{ const.}\right\} \\
&= \dot{x}\bar{e}_x + \dot{y}\bar{e}_y + \dot{z}\bar{e}_z. \qquad\qquad \square
\end{aligned}$$

The time derivatives of the basis vectors are zero, since they are constants for a rectangular coordinate system.

Theorem 7.5 (Acceleration in rectangular coordinates). In a rectangular coordinate system, the acceleration of a particle with position vector $\bar{r} = x\bar{e}_x + y\bar{e}_y + z\bar{e}_z$ is

$$\bar{a} = \ddot{x}\bar{e}_x + \ddot{y}\bar{e}_y + \ddot{z}\bar{e}_z. \tag{7.9}$$

Proof. Definition 7.3 of acceleration gives

$$\begin{aligned}
\bar{a} &= \frac{d\bar{v}}{dt} = \left\{\text{Theorem 7.4}\right\} \\
&= \frac{d}{dt}\left(\dot{x}\bar{e}_x + \dot{y}\bar{e}_y + \dot{z}\bar{e}_z\right) = \left\{\text{Product rule (A.25a)}\right\} \\
&= \ddot{x}\bar{e}_x + \dot{x}\frac{d\bar{e}_x}{dt} + \ddot{y}\bar{e}_y + \dot{y}\frac{d\bar{e}_y}{dt} + \ddot{z}\bar{e}_z + \dot{z}\frac{d\bar{e}_z}{dt} = \left\{\bar{e}_x, \bar{e}_y, \bar{e}_z \text{ const.}\right\} \\
&= \ddot{x}\bar{e}_x + \ddot{y}\bar{e}_y + \ddot{z}\bar{e}_z. \qquad\qquad \square
\end{aligned}$$

Similarly to rectilinear motion, it is desirable to rewrite the expressions for velocity and acceleration on differential form, so that the particle motion can be obtained by integration.

Theorem 7.6. If the path of a particle is $\bar{r} = x\bar{e}_x + y\bar{e}_y + z\bar{e}_z$, its velocity is $\bar{v} = v_x\bar{e}_x + v_y\bar{e}_y + v_z\bar{e}_z$, and its acceleration is $\bar{a} = a_x\bar{e}_x + a_y\bar{e}_y + a_z\bar{e}_z$, then the following differential relations hold

$$\begin{aligned}
dx &= v_x dt, & dy &= v_y dt, & dz &= v_z dt, \\
dv_x &= a_x dt, & dv_y &= a_y dt, & dv_z &= a_z dt, \\
v_x dv_x &= a_x dx, & v_y dv_y &= a_y dy, & v_z dv_z &= a_z dz.
\end{aligned} \tag{7.10}$$

Proof. For coordinate direction x, Eq. (7.8) gives

$$v_x = \frac{\mathrm{d}x}{\mathrm{d}t} \quad \Leftrightarrow \quad \left\{ \text{Eq. (A.30)} \right\} \quad \Leftrightarrow$$
$$dx = v_x dt. \tag{7.11}$$

Furthermore, Eq. (7.9) gives

$$a_x = \frac{\mathrm{d}^2 x}{\mathrm{d}t^2} = \frac{\mathrm{d}v_x}{\mathrm{d}t} \quad \Leftrightarrow \quad \left\{ \text{Eq. (A.30)} \right\} \quad \Leftrightarrow$$
$$dv_x = a_x dt.$$

By multiplying this equation by v_x, we obtain

$$v_x dv_x = a_x v_x dt \quad \Leftrightarrow \quad \left\{ \text{Eq. (7.11)} \right\} \quad \Leftrightarrow$$
$$v_x dv_x = a_x dx.$$

The corresponding differential relations for coordinate directions y and z are obtained analogously. $\qquad\square$

According to Theorem 7.6, the differential relations governing rectilinear motion hold for each coordinate direction during curvilinear motion.

Polar coordinates

Consider a rectangular coordinate system with origin $\mathcal{O}$, and a particle $\mathcal{P}$ in the xy plane. The position of this particle can be uniquely represented by the distance $r = |\overline{\mathcal{OP}}|$ and the counterclockwise angle θ from the x axis to the ray $\mathcal{OP}$. We define the *polar coordinates* of $\mathcal{P}$ to be r and θ (Fig. 7.5). If the particle moves in the xy plane, its polar coordinates are time-dependent: $r = r(t)$ and $\theta = \theta(t)$. We define the *angular velocity* of the particle as

$$\omega \equiv \dot{\theta}, \tag{7.12}$$

and we define the *angular acceleration* of the particle as

$$\alpha \equiv \dot{\omega} = \ddot{\theta}. \tag{7.13}$$

Definition 7.7 (Polar basis vectors). Given a rectangular basis $\{\bar{e}_x, \bar{e}_y\}$, the *polar basis vectors* are defined as (Fig. 7.6)

$$\bar{e}_r \equiv \cos\theta\, \bar{e}_x + \sin\theta\, \bar{e}_y, \tag{7.14a}$$
$$\bar{e}_\theta \equiv -\sin\theta\, \bar{e}_x + \cos\theta\, \bar{e}_y. \tag{7.14b}$$

Using Def. 7.7, the position vector can be written in polar form as

$$\bar{r} = r\bar{e}_r, \tag{7.15}$$

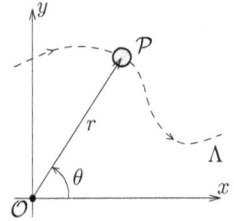

Figure 7.5: Polar coordinates (r, θ).

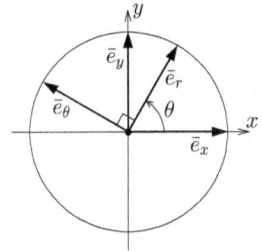

Figure 7.6: Rectangular and polar basis vectors in the unit circle.

where $\bar{r} = \bar{r}(t)$, $r = r(t)$ and $\bar{e}_r = \bar{e}_r(t)$. The velocity and the acceleration of a particle can be determined from Defs. 7.2 and 7.3, respectively, by time differentiation of Eq. (7.15). This differentiation is simplified, if we first calculate the time derivatives of the polar basis vectors.

Theorem 7.8. The time derivatives of the polar basis vectors are

$$\frac{\mathrm{d}\bar{e}_r}{\mathrm{d}t} = \dot{\theta}\bar{e}_\theta, \tag{7.16a}$$

$$\frac{\mathrm{d}\bar{e}_\theta}{\mathrm{d}t} = -\dot{\theta}\bar{e}_r. \tag{7.16b}$$

Proof. Differentiation of Eq. (7.14a) gives

$$\begin{aligned}
\frac{\mathrm{d}\bar{e}_r}{\mathrm{d}t} &= \frac{\mathrm{d}}{\mathrm{d}t}(\cos\theta\bar{e}_x) + \frac{\mathrm{d}}{\mathrm{d}t}(\sin\theta\bar{e}_y) = \{\bar{e}_x, \bar{e}_y \text{ const.}\} \\
&= \frac{\mathrm{d}(\cos\theta)}{\mathrm{d}t}\bar{e}_x + \frac{\mathrm{d}(\sin\theta)}{\mathrm{d}t}\bar{e}_y = \{\text{Chain rule}\} \\
&= \frac{\mathrm{d}(\cos\theta)}{\mathrm{d}\theta}\frac{\mathrm{d}\theta}{\mathrm{d}t}\bar{e}_x + \frac{\mathrm{d}(\sin\theta)}{\mathrm{d}\theta}\frac{\mathrm{d}\theta}{\mathrm{d}t}\bar{e}_y \\
&= (-\sin\theta)\dot{\theta}\bar{e}_x + (\cos\theta)\dot{\theta}\bar{e}_y \\
&= \dot{\theta}(-\sin\theta\bar{e}_x + \cos\theta\bar{e}_y) = \{\text{Eq. (7.14b)}\} \\
&= \dot{\theta}\bar{e}_\theta.
\end{aligned}$$

Differentiation of Eq. (7.14b) gives

$$\begin{aligned}
\frac{\mathrm{d}\bar{e}_\theta}{\mathrm{d}t} &= -\frac{\mathrm{d}}{\mathrm{d}t}(\sin\theta\bar{e}_x) + \frac{\mathrm{d}}{\mathrm{d}t}(\cos\theta\bar{e}_y) = \{\bar{e}_x, \bar{e}_y \text{ const.}\} \\
&= -\frac{\mathrm{d}(\sin\theta)}{\mathrm{d}t}\bar{e}_x + \frac{\mathrm{d}(\cos\theta)}{\mathrm{d}t}\bar{e}_y = \{\text{Chain rule}\} \\
&= -\frac{\mathrm{d}(\sin\theta)}{\mathrm{d}\theta}\frac{\mathrm{d}\theta}{\mathrm{d}t}\bar{e}_x + \frac{\mathrm{d}(\cos\theta)}{\mathrm{d}\theta}\frac{\mathrm{d}\theta}{\mathrm{d}t}\bar{e}_y \\
&= -(\cos\theta)\dot{\theta}\bar{e}_x + (-\sin\theta)\dot{\theta}\bar{e}_y \\
&= -\dot{\theta}(\cos\theta\bar{e}_x + \sin\theta\bar{e}_y) = \{\text{Eq. (7.14a)}\} \\
&= -\dot{\theta}\bar{e}_r. \qquad \square
\end{aligned}$$

Straight-forward time differentiation of Eq. (7.15) then gives the expressions for velocity and acceleration in polar coordinates.

Theorem 7.9 (Velocity in polar coordinates). In polar form, the velocity of a particle is

$$\bar{v} = \dot{r}\bar{e}_r + r\dot{\theta}\bar{e}_\theta. \tag{7.17}$$

Proof. From Def. 7.2 of velocity, we have

$$\begin{aligned}
\bar{v} &= \frac{\mathrm{d}\bar{r}}{\mathrm{d}t} = \{\text{Eq. (7.15)}\} \\
&= \frac{\mathrm{d}}{\mathrm{d}t}(r\bar{e}_r) = \{\text{Product rule (A.25a)}\}
\end{aligned}$$

$$= \dot{r}\bar{e}_r + r\frac{\mathrm{d}\bar{e}_r}{\mathrm{d}t} = \{\text{Eq. (7.16a)}\}$$

$$= \dot{r}\bar{e}_r + r\dot{\theta}\bar{e}_\theta. \qquad \qquad \square$$

Theorem 7.10 (Acceleration in polar coordinates). In polar form, the acceleration of a particle is

$$\bar{a} = (\ddot{r} - r\dot{\theta}^2)\bar{e}_r + (r\ddot{\theta} + 2\dot{r}\dot{\theta})\bar{e}_\theta. \qquad (7.18)$$

Proof. From Def. 7.3 of acceleration, we have

$$\bar{a} = \frac{\mathrm{d}\bar{v}}{\mathrm{d}t} = \{\text{Eq. (7.17)}\}$$

$$= \frac{\mathrm{d}}{\mathrm{d}t}(\dot{r}\bar{e}_r + r\dot{\theta}\bar{e}_\theta) = \{\text{Product rule (A.25a)}\}$$

$$= \ddot{r}\bar{e}_r + \dot{r}\frac{\mathrm{d}\bar{e}_r}{\mathrm{d}t} + \dot{r}\dot{\theta}\bar{e}_\theta + r\ddot{\theta}\bar{e}_\theta + r\dot{\theta}\frac{\mathrm{d}\bar{e}_\theta}{\mathrm{d}t} = \{\text{Eqs. (7.16a), (7.16b)}\}$$

$$= \ddot{r}\bar{e}_r + \dot{r}(\dot{\theta}\bar{e}_\theta) + \dot{r}\dot{\theta}\bar{e}_\theta + r\ddot{\theta}\bar{e}_\theta + r\dot{\theta}(-\dot{\theta}\bar{e}_r)$$

$$= (\ddot{r} - r\dot{\theta}^2)\bar{e}_r + (r\ddot{\theta} + 2\dot{r}\dot{\theta})\bar{e}_\theta. \qquad \square$$

Circular motion

Circular motion is conveniently described using polar coordinates (Fig. 7.7). By placing the origin in the center of the circular path of radius R, we ensure that $r = R$ is constant so that $\dot{r} = 0$ and $\ddot{r} = 0$. Therefore, the expressions for velocity and acceleration simplify to

$$\bar{v} = r\dot{\theta}\bar{e}_\theta = R\omega\bar{e}_\theta, \qquad (7.19a)$$

$$\bar{a} = -r\dot{\theta}^2\bar{e}_r + r\ddot{\theta}\bar{e}_\theta = -R\omega^2\bar{e}_r + R\alpha\bar{e}_\theta, \qquad (7.19b)$$

for circular motion. From Eq. (7.19a), we find a simple relation between the speed of the particle and the angular velocity:

$$v = R|\omega|. \qquad (7.20)$$

However, this expression only holds for circular motion.

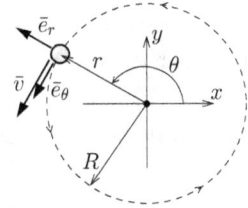

Figure 7.7: Circular motion in a polar coordinate system.

Arc coordinates

Consider a particle $\mathcal{P}$ that moves along a path Λ in the plane. From a fixed point $\mathcal{O}$ on this path, the position vector of the particle can be written as $\bar{r} = \bar{r}(s)$, where $s = s(t)$ is the *arc coordinate*, being the arc length from $\mathcal{O}$ to $\mathcal{P}$ along the path (Fig. 7.8).

Definition 7.11 (Natural basis vectors). For a given arc parametrization $\bar{r} = \bar{r}(s)$, the natural basis vectors are

$$\bar{e}_t \equiv \frac{\mathrm{d}\bar{r}}{\mathrm{d}s}, \qquad (7.21a)$$

$$\bar{e}_n \equiv \rho\frac{\mathrm{d}\bar{e}_t}{\mathrm{d}s}, \qquad \rho \equiv \left|\frac{\mathrm{d}\bar{e}_t}{\mathrm{d}s}\right|^{-1}, \qquad (7.21b)$$

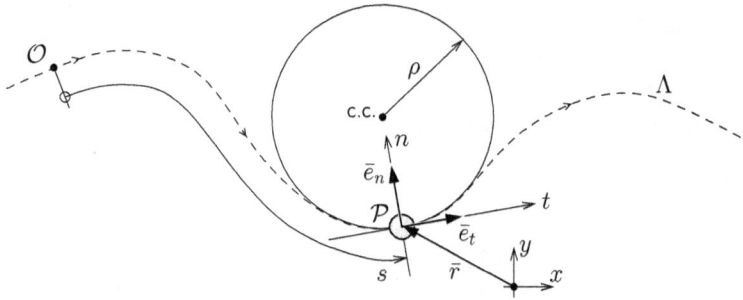

Figure 7.8: The motion of a particle $\mathcal{P}$ along a path Λ in the plane, having the arc coordinate s as measured from the fixed point $\mathcal{O}$. The natural basis vectors $\bar{e}_t$ and $\bar{e}_n$ vary with the position of the particle.

where $\bar{e}_t$ is the *tangent unit vector* of the path, $\bar{e}_n$ is the *normal unit vector*, and ρ is the *radius of curvature* of the path.

Definition 7.11 is formulated so that $|\bar{e}_t| = |\bar{e}_n| = 1$ and $\bar{e}_t \perp \bar{e}_n$. Thus, the unit vectors form an orthonormal basis in the plane. The directions of the basis vectors vary with the position of the particle (Fig. 7.8).

For our purposes, a geometric interpretation of Def. 7.11 is of interest. When the particle $\mathcal{P}$ is in a given position, the tangent unit vector $\bar{e}_t$ points in the positive direction of the arc coordinate. Moreover, it is possible to construct an *osculating circle*, which is tangent to the path at $\mathcal{P}$, and has the same radius of curvature ρ as the path at $\mathcal{P}$ (Fig. 7.8). The center of this osculating circle is the *center of curvature* (c.c.), and the normal unit vector is directed from the particle towards the center of curvature.

Since the position of a particle varies with time t, we write $s = s(t)$. Thus, the position vector of the particle becomes

$$\bar{r} = \bar{r}\,[s(t)]\,, \quad \dot{s} \geq 0. \tag{7.22}$$

The condition $\dot{s} \geq 0$ implies that s represents the *distance* covered by the particle. As a consequence, a particle that moves back and forth along the same path, as the pendulum in Fig. 7.9, must be attributed a folding path. Expressions for the velocity and the acceleration are obtained by differentiating Eq. (7.22) w.r.t. time.

Theorem 7.12 (Velocity in the natural basis). In the natural basis, the velocity of a particle is

$$\bar{v} = \dot{s}\bar{e}_t = v\bar{e}_t, \tag{7.23}$$

where $v = \dot{s} \geq 0$ is the speed of the particle (Fig. 7.10).

Proof. From Def. 7.2 of velocity, we obtain

$$\bar{v} = \frac{d\bar{r}}{dt} = \{\text{Eq. (7.22)}\}$$

$$= \frac{d}{dt}\bar{r}\,[s(t)] = \{\text{Chain rule}\}$$

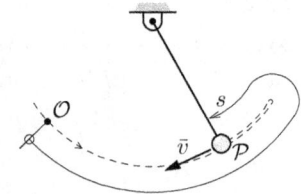

Figure 7.9: The motion of a pendulum back and forth in a vertical plane is represented by a folding path, so that the arc coordinate increases with time.

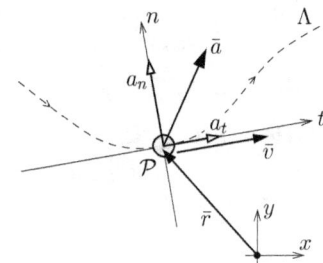

Figure 7.10: Position vector $\bar{r}$, velocity $\bar{v}$, and acceleration $\bar{a}$ in the natural basis.

$$= \frac{d\bar{r}}{ds}\frac{ds}{dt} = \big\{\text{Eq. (7.21a)}\big\}$$

$$= \dot{s}\bar{e}_t. \qquad \qquad \square$$

Theorem 7.13 (Acceleration in the natural basis). In the natural basis, the acceleration of a particle is

$$\bar{a} = \dot{v}\bar{e}_t + \frac{v^2}{\rho}\bar{e}_n, \qquad (7.24)$$

where $v = \dot{s}$, and ρ is the radius of curvature of the path of motion.

Proof. From Def. 7.3 of acceleration, we have

$$\bar{a} = \frac{d\bar{v}}{dt} = \big\{\text{Eq. (7.23)}\big\}$$

$$= \frac{d}{dt}(\dot{s}\bar{e}_t) = \big\{\text{Product rule (A.25a)}\big\}$$

$$= \ddot{s}\bar{e}_t + \dot{s}\frac{d\bar{e}_t}{dt} = \big\{\text{Chain rule}\big\}$$

$$= \ddot{s}\bar{e}_t + \dot{s}\frac{d\bar{e}_t}{ds}\frac{ds}{dt} = \big\{\text{Eq. (7.21b)}\big\}$$

$$= \dot{v}\bar{e}_t + \frac{v^2}{\rho}\bar{e}_n. \qquad \qquad \square$$

As a consequence of Theorem 7.13, the normal component of the acceleration is $a_n = v^2/\rho$, and it is always directed towards the center of curvature (Fig. 7.10).

Theorem 7.14. For an arc parametrization of the motion of a particle, it holds that

$$vdv = a_t ds, \qquad (7.25)$$

where s is the arc coordinate, v is the speed, and $a_t = \dot{v}$ is the tangential component of the acceleration.

Proof. Since $v = ds/dt$, Eq. (A.30) gives

$$ds = vdt. \qquad (7.26)$$

According to Eq. (A.29), we have

$$dv = \dot{v}dt \quad \Leftrightarrow \quad \big\{\text{Eq. (7.24)}\big\} \quad \Leftrightarrow$$

$$dv = a_t dt \quad \Leftrightarrow$$

$$vdv = a_t vdt \quad \Leftrightarrow \quad \big\{\text{Eq. (7.26)}\big\} \quad \Leftrightarrow$$

$$vdv = a_t ds. \qquad \qquad \square$$

7.3 Kinematic constraints

When two particles are mechanically connected by, *e.g.*, a stretched string or a link, their motions are interlinked by a *kinematic constraint*. Here, we study two examples of how constraints are enforced.

Particles connected by a string

Consider two particles, $\mathcal{P}$ and $\mathcal{Q}$, connected by a string as illustrated in Fig. 7.11. This string runs over two pulleys, each one with radius R. Using the lengths and positions defined in Fig. 7.11, we express the length of the string as

$$
\begin{aligned}
\ell &= x_{\mathcal{P}} + \pi R + (x_{\mathcal{P}} - d) + \pi R + x_{\mathcal{Q}} \\
&= 2x_{\mathcal{P}} + x_{\mathcal{Q}} + 2\pi R - d.
\end{aligned} \tag{7.27}
$$

Two differentiations of Eq. (7.27) w.r.t. time yield relations between the velocities and accelerations of the particles:

$$
\begin{aligned}
0 &= 2\dot{x}_{\mathcal{P}} + \dot{x}_{\mathcal{Q}} &&\Leftrightarrow&& 2v_{\mathcal{P}} + v_{\mathcal{Q}} = 0, \\
0 &= 2\ddot{x}_{\mathcal{P}} + \ddot{x}_{\mathcal{Q}} &&\Leftrightarrow&& 2a_{\mathcal{P}} + a_{\mathcal{Q}} = 0,
\end{aligned}
$$

where we used that the length of the string is constant. The relation between the accelerations of different particles is often needed, since the acceleration appears in the Law of force and acceleration (Sect. 1.2).

Generally, it is instrumental to express the length of a string, using the coordinates of the connected particles, and subsequently differentiate w.r.t. time. If the problem involves several strings, this procedure should be applied for each string.

Particles connected by a link

For two particles connected by a rotating link, we introduce an angular coordinate θ, that represents the turning of the link relative to a fixed axis. If this angle remains small, the motion of the endpoints of the link is approximately rectilinear.

Consider two particles, $\mathcal{P}$ and $\mathcal{Q}$, that are suspended in each endpoint of a straight bar (Fig. 7.12). The positions of the particles can be written as functions of the turning angle θ:

$$
\begin{cases} x_{\mathcal{P}} = b + \ell_{\mathcal{P}} \sin\theta \\ x_{\mathcal{Q}} = b - \ell_{\mathcal{Q}} \sin\theta \end{cases} \Rightarrow \quad x_{\mathcal{P}} = b + \frac{\ell_{\mathcal{P}}}{\ell_{\mathcal{Q}}}(b - x_{\mathcal{Q}}),
$$

where $\sin\theta$ was eliminated from the system of equations. Two subsequent differentiations w.r.t. time give

$$
\begin{aligned}
\dot{x}_{\mathcal{P}} &= -\frac{\ell_{\mathcal{P}}}{\ell_{\mathcal{Q}}}\dot{x}_{\mathcal{Q}} &&\Leftrightarrow&& v_{\mathcal{P}} = -\frac{\ell_{\mathcal{P}}}{\ell_{\mathcal{Q}}}v_{\mathcal{Q}}, \\
\ddot{x}_{\mathcal{P}} &= -\frac{\ell_{\mathcal{P}}}{\ell_{\mathcal{Q}}}\ddot{x}_{\mathcal{Q}} &&\Leftrightarrow&& a_{\mathcal{P}} = -\frac{\ell_{\mathcal{P}}}{\ell_{\mathcal{Q}}}a_{\mathcal{Q}}.
\end{aligned}
$$

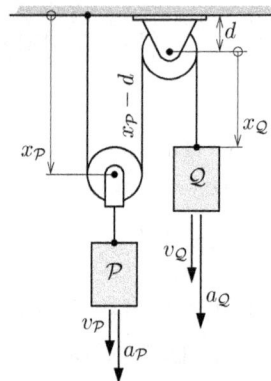

Figure 7.11: Two particles $\mathcal{P}$ and $\mathcal{Q}$ connected by a string, that runs over pulleys with radii R.

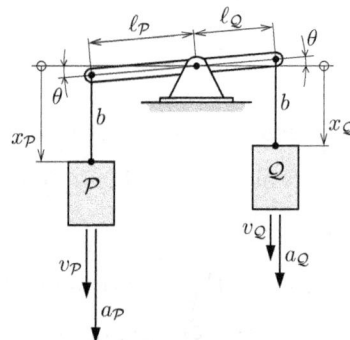

Figure 7.12: Two particles, $\mathcal{P}$ and $\mathcal{Q}$, suspended in strings and connected by a link.

8
Kinetics of particles

8.1 Newton's laws of motion

We repeat Newton's laws of motion for particles, *cf.* Sect. 1.2, which constitute the foundation of particle kinetics:

Postulate 8.1 (Law of inertia). A particle remains at rest, or moves in a straight line at a constant velocity, as long as the particle does not interact with any other objects.

Postulate 8.2 (Law of force and acceleration). For a particle with constant mass m, it holds that

$$\Sigma \bar{F} = m\bar{a}, \tag{8.1}$$

where $\Sigma \bar{F}$ is the vector sum of the forces acting on the particle, and $\bar{a}$ is the acceleration of the particle.

Postulate 8.3 (Law of action and reaction). If a particle $\mathcal{P}$ exerts a force $\bar{F}$ on another particle $\mathcal{Q}$, then $\mathcal{Q}$ exerts a force $-\bar{F}$ on $\mathcal{P}$. That is, the forces of action and reaction between two particles have equal magnitude, but opposite directions.

The postulates above are called Newton's first, second and third law, respectively. Experiments show that Newton's laws of motion hold for macroscopic systems. That is, for systems much larger than the atomic length scale, and for speeds much lower than the speed of light.

Newton's first law

According to Newton's first law, Postulate 8.1, no forces are required to maintain motion. A particle moves at a constant velocity, in so-called *uniform motion*, if it does not interact with any surrounding objects; some sort of interaction is required to change its uniform motion. This resistance of a particle against changing its velocity is called *inertia*.

To describe the motion of a particle, it is necessary to introduce a geometric reference system whose position, orientation and scale in space are defined in relation to physical objects. These systems are called *frames of reference*. A coordinate system is introduced in the chosen frame of reference to quantify the positions, velocities and accelerations of particles.

Newton's laws of particle motion are valid only in certain frames of reference known as *inertial frames of reference*. The corresponding coordinate systems are called inertial systems. Newton's first law, the Law of inertia, makes it possible to determine whether a given coordinate system is an inertial system. To make this distinction, one chooses a number of objects, which are interacting weakly with their surroundings, *e.g.*, distant stars isolated from other massive celestial bodies. If each such object has a constant velocity in the given coordinate system (Fig. 8.1a), then that coordinate system is indeed an inertial system. Conversely, if the velocities of these isolated objects vary in the given coordinate system (Fig. 8.1b), then this cannot be an inertial system.

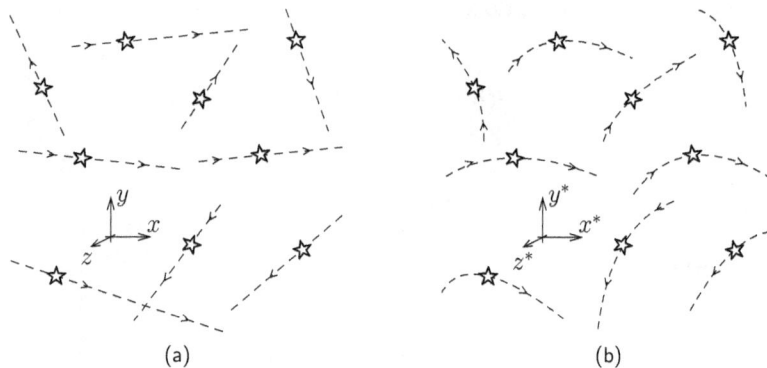

Figure 8.1: (a) Particles with negligible interaction describe uniform motion in the coordinate system xyz. Hence, xyz is an inertial system. (b) Essentially isolated objects appear to accelerate in the coordinate system $x^*y^*z^*$. Hence, $x^*y^*z^*$ is *not* an inertial system.

A coordinate system that is fixed relative to the surface of the Earth is not strictly an inertial system. This becomes apparent when one photographs a clear night sky using a long time of exposure (Fig. 8.2); the stars will not describe uniform motion w.r.t. the terrestrial frame of reference. However, in many applications—not in all—sufficient accuracy is achieved by applying Newton's laws for a terrestrial coordinate system.

Newton's second law

Newton's second law, the Law of force and acceleration, is only valid in inertial frames of reference, as defined by the Law of inertia. By selecting an inertial system, the acceleration $\bar{a}$ in the Law of force and acceleration becomes well-defined (Def. 7.3).

In more physically-oriented presentations, the concept of mass is de-

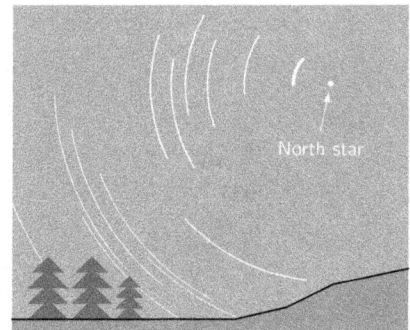

Figure 8.2: Long time exposure of the night sky as seen from the Earth. A terrestrial system is not an inertial system.

fined using Newton's laws.[11] Herein, however, it is presumed that mass and force are two previously well-defined quantities. Their relation to the motion of a particle is given by the Law of force and acceleration:

$$\Sigma \bar{F} = m\bar{a}.$$

The left-hand side of this equation contains the vector sum of all forces acting on the particle. Note, however, that only forces arising from interactions, for instance forces of gravity and contact forces, are to be included in this sum.[12] The mass m quantifies the resistance of the particle to changing its velocity. That is, mass quantifies inertia.

Newton's third law

Newton's third law, the Law of action and reaction for particles, describes the reciprocal nature of interactions. Since forces arise through interaction between particles, the forces appear in pairs. The force and the reaction force are equal in magnitude, but have opposite directions (Fig. 8.3). The third law does not hold forth whether the action and reaction create a couple. Therefore, we formulate an amendment to the Law of action and reaction:

Postulate 8.4. For interactions between particles, the force and its reaction force share a common line of action (Fig. 8.3).

The Law of action and reaction can be used in both Statics and Dynamics, and it holds even for deformable bodies. However, there are cases when this law fails, for instance, when accelerating particles interact through electromagnetic forces, or when the distance between particles is extremely large.[13]

8.2   Equations of motion and problem solving

In kinetics problems, the motion of a particle is determined by the forces that act on this particle.

Rectilinear motion

For rectilinear motion the particle moves along a straight line in an inertial system. Herein, we choose a rectangular coordinate system, such that the x direction coincides with the direction of motion. Since there is no motion in the y or z directions, it holds that $a_y = a_z = 0$. For this reason, the component form of the Law of force and acceleration

[11] J. B. Griffiths. *The theory of classical mechanics*. Cambridge University Press, 1985. ISBN 0-521-23760-2

[12] Fictitious forces, *e.g.* centrifugal forces, do not obey the laws of proper forces, such as the Law of action and reaction.

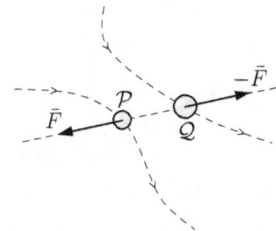

Figure 8.3: Newton's third law, the Law of action and reaction, subject to the additional supposition, that the force and its reaction have a common line of action.

[13] K. R. Symon. *Mechanics*. Addison-Wesley Publishing Company, Inc., 2nd edition, 1960

becomes

$$\Sigma F_x = ma_x = m\ddot{x}, \tag{8.2a}$$

$$\Sigma F_y = 0, \tag{8.2b}$$

$$\Sigma F_z = 0, \tag{8.2c}$$

where we used Theorem 7.5. Thus, the acceleration in the given direction of motion is determined by the mass of the particle and the force sum in this direction.

Curvilinear planar motion

When the motion of a particle is restricted to a plane, there are three alternative coordinate systems (Chapter 7), which can be used to describe this motion.

For rectangular coordinates, we align, *e.g.*, the xy plane with the plane of motion. Then, Theorem 7.5 states that $a_x = \ddot{x}$, $a_y = \ddot{y}$ and $a_z = 0$, and the component form of the Law of force and acceleration becomes

$$\Sigma F_x = ma_x = m\ddot{x}, \tag{8.3a}$$

$$\Sigma F_y = ma_y = m\ddot{y}, \tag{8.3b}$$

$$\Sigma F_z = ma_z = 0. \tag{8.3c}$$

Similarly, for polar coordinates (r, θ) and planar motion, Theorem 7.10 ensures that $a_r = \ddot{r} - r\dot{\theta}^2$ and $a_\theta = r\ddot{\theta} + 2\dot{r}\dot{\theta}$. Thus, the component form of the Law of force and acceleration becomes

$$\Sigma F_r = ma_r = m(\ddot{r} - r\dot{\theta}^2), \tag{8.4a}$$

$$\Sigma F_\theta = ma_\theta = m(r\ddot{\theta} + 2\dot{r}\dot{\theta}). \tag{8.4b}$$

These equations are simplified considerably for the case of circular motion, for which $\dot{r} = 0$ and $\ddot{r} = 0$ if the origin is placed in the center of the circular path of motion.

For arc coordinates and planar motion, Theorem 7.13 states that $a_t = \dot{v}$ and $a_n = v^2/\rho$, where ρ is the radius of curvature of the path. Thus, the component form of the Law of force and acceleration becomes

$$\Sigma F_n = ma_n = m\frac{v^2}{\rho}, \tag{8.5a}$$

$$\Sigma F_t = ma_t = m\dot{v}. \tag{8.5b}$$

It is of critical importance to apply the correct coordinate directions. For arc coordinates, the normal direction must be towards the center of the curvature.

(a)

(b)

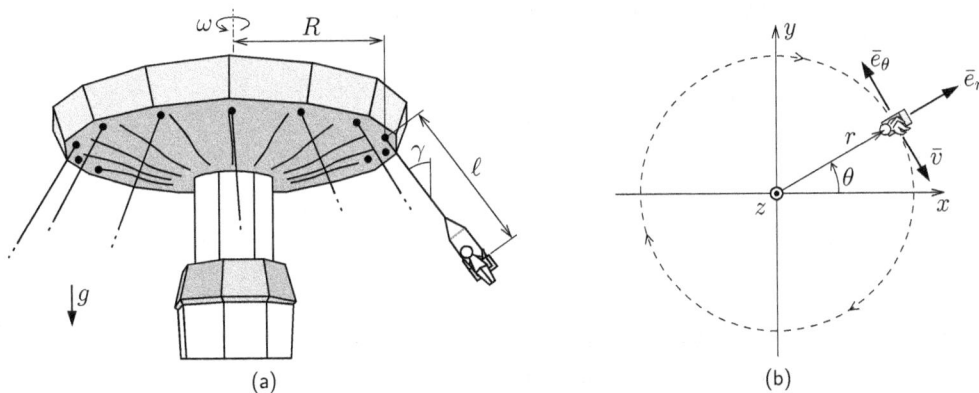

Figure 8.4: (a) A swingride rotates with constant angular velocity ω. (b) A chair with a passenger is considered as one particle, which describes circular motion in a horizontal plane.

Planar motion with three-dimensional force-couple systems

Although the motion is confined to a plane, it is sometimes necessary to consider the balance of the out-of-plane force components. We analyze an example to illustrate this problem class.

Consider the swingride in Fig. 8.4a. A horizontal, circular roof with radius R is rotating about a vertical axis at a constant angular velocity ω. Massless chains with length ℓ are attached at the outer edge of this roof. In the other end of each chain, there is a chair with a passenger of mass m. During the ride, each chain forms a constant angle $\gamma \in [0, \frac{\pi}{2}[$ with a vertical line. Determine the corresponding angular velocity ω and the tensile force T in the chains.

Choice of coordinate system: We consider the chair with a passenger as one particle, which describes circular motion in a horizontal plane (Fig. 8.4b). We introduce a polar coordinate system in the plane of motion, so that

$$r = R + \ell \sin \gamma, \quad \dot{r} = 0, \quad \ddot{r} = 0, \quad \dot{\theta} = -\omega, \quad \ddot{\theta} = 0.$$

Free-body diagram: We introduce a third coordinate z, perpendicular to the plane of motion, and we draw a free-body diagram for the particle (the chair with a passenger) in the rz plane:

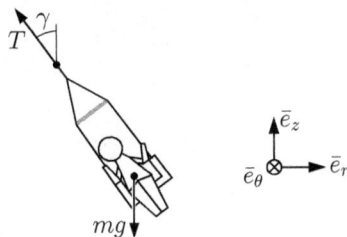

Kinematics relations: According to Theorem 7.10, and the fact that there is no motion in the z direction, we have

$$a_r = \ddot{r} - r\dot{\theta}^2 = -(R + \ell\sin\gamma)\omega^2,$$
$$a_\theta = r\ddot{\theta} + 2\dot{r}\dot{\theta} = 0,$$
$$a_z = 0.$$

Law of force and acceleration: We express the Law of force and acceleration, $\Sigma\bar{F} = m\bar{a}$, for the particle on component form:

$$\rightarrow^r: \qquad -T\sin\gamma = ma_r = -m(R + \ell\sin\gamma)\omega^2, \qquad (8.7\text{a})$$
$$\uparrow^z: \qquad T\cos\gamma - mg = ma_z = 0. \qquad (8.7\text{b})$$

Calculations: Equations (8.7a) and (8.7b) form a system of equations, which is solved for the two unknowns, ω and T, giving

$$T = \frac{mg}{\cos\gamma},$$
$$\omega = \pm\sqrt{\frac{g\tan\gamma}{R + \ell\sin\gamma}}.$$

The '$\pm$' symbol in front of ω indicates that both directions of rotation result in the same constant angle γ. $\qquad\qquad\square$

The problem-solving scheme above can be adapted and used in a range of kinetics applications.

9

Work–energy method for particles

When the forces acting on a particle can be written as functions of its position (Fig. 9.1), the analysis may be simplified by using the *Work–energy method*, which considers the transformation of work into kinetic energy and *vice versa*. Herein, we choose to define work based on the concept of *power*. Henceforth, it is presumed that an inertial system is chosen to describe motion.

Figure 9.1: When the forces on a particle $\mathcal{P}$ depend on its position, the Work–energy method is often useful.

9.1 Power and Work

Definition 9.1 (Power of a force). The *power* developed by a force $\bar{F}$ is

$$P \equiv \bar{F} \cdot \bar{v}, \tag{9.1}$$

where $\bar{v}$ is the velocity of the point of action of the force.

From Eq. (9.1), it is clear that a force with a fixed point of action, $\bar{v} = \bar{0}$, does not develop any power. The power is measured in the SI unit of watt (W), or in derived USC units:

$$1\,\mathrm{W} = 1\,\frac{\mathrm{kg \cdot m^2}}{\mathrm{s^3}}, \qquad 1\,\frac{\mathrm{lb_f \cdot ft}}{\mathrm{s}} = \frac{1}{550}\,\mathrm{hp} \approx 1.3558\,\mathrm{W},$$

where we introduced the *mechanical horsepower* (hp).

9.2 Work

Work and energy are measured in the SI unit of joule (J), or in the USC units of $\mathrm{lb_f \cdot ft}$, where

$$1\,\mathrm{J} = 1\,\mathrm{N \cdot m} = 1\,\frac{\mathrm{kg \cdot m^2}}{\mathrm{s^2}}, \qquad 1\,\mathrm{lb_f \cdot ft} \approx 1.3558\,\mathrm{J}.$$

Definition 9.2 (Work of a force). The *work* of a force $\bar{F}$, between times t_1 and t_2, is

$$U_{1-2} \equiv \int_{t_1}^{t_2} P\mathrm{d}t = \int_{t_1}^{t_2} \bar{F} \cdot \bar{v}\mathrm{d}t, \tag{9.2}$$

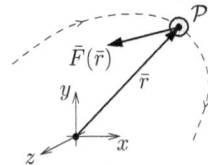

where P is the power of the force, and $\bar{v}$ is the velocity of the point of action of the force.

When solving problems, one can rewrite the integral over time in Eq. (9.2), into an integral along the path of the point of action, or equivalently along the path of the particle under analysis.

Theorem 9.3 (Work integral). Let $\bar{r} = \bar{r}(s)$ be the path of a particle, where s is the arc coordinate.[14] The work of a force $\bar{F}(s)$ acting on this particle between positions 1 and 2 is

[14] Since $\dot{s} \geq 0$, $\bar{r}(s)$ represents a *unique* path, and s is the traversed distance.

$$U_{1-2} = \int_{s_1}^{s_2} \bar{F} \cdot \bar{e}_t \mathrm{d}s, \qquad (9.3)$$

where $\bar{e}_t$ is the tangent direction of the path, and s_1 and s_2 denote the arc coordinates of positions 1 and 2, respectively (Fig. 9.2).

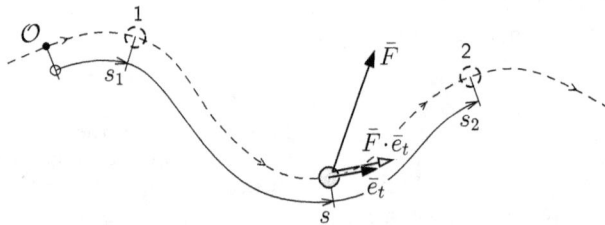

Figure 9.2: The path of a particle between arc coordinate positions s_1 and s_2.

Proof. Let t_1 and t_2 be the times corresponding to particle positions 1 and 2, respectively. Then, according to Def. (9.2), we have

$$
\begin{aligned}
U_{1-2} &= \int_{t_1}^{t_2} \bar{F} \cdot \bar{v} \mathrm{d}t = \big\{ \text{Eq. (7.23)} \big\} \\
&= \int_{t_1}^{t_2} \bar{F}[s(t)] \cdot \dot{s}\bar{e}_t \mathrm{d}t = \left\{ \begin{array}{l} \text{subst. (A.39)} \\ s = s(t),\ ds = \frac{\mathrm{d}s}{\mathrm{d}t}dt \end{array} \right\} \\
&= \int_{s(t_1)}^{s(t_2)} \bar{F} \cdot \bar{e}_t \mathrm{d}s. \qquad \square
\end{aligned}
$$

A trivial, yet important, consequence of Theorem 9.3 is that no work is done by a force acting in a space-fixed[15] point such that $s_1 = s_2$.

[15] *Space-fixed* – with unchanging geometry in a given inertial system.

Definition 9.4 (Work on a particle). The work done on a particle between times t_1 and t_2 is

$$\Sigma U_{1-2} = \int_{t_1}^{t_2} \Sigma \bar{F} \cdot \bar{v} \mathrm{d}t, \qquad (9.4)$$

where $\bar{v}$ is the velocity of the particle, and $\Sigma \bar{F}$ is the force sum acting on the particle.

Thus, the total work done on a particle between times t_1 and t_2 is

$$\Sigma U_{1-2} = \int_{t_1}^{t_2} \Sigma \bar{F} \cdot \bar{v} \mathrm{d}t = \int_{t_1}^{t_2} \left(\sum_{i=1}^{n} \bar{F}_i \cdot \bar{v} \right) \mathrm{d}t = \sum_{i=1}^{n} \int_{t_1}^{t_2} \bar{F}_i \cdot \bar{v} \mathrm{d}t. \quad (9.5)$$

That is, the total work is the sum of the work of each force.

Constraint forces, for instance the normal force $\bar{N}$, only arise in directions in which particle motion is restricted. Therefore, the motion of a particle relative to a space-fixed obstruction is always perpendicular to the constraint force: $\bar{v} \perp \bar{N}$. As a consequence, since $\bar{N} \cdot \bar{v} = 0$, constraint forces from such obstructions neither yield power, nor work (Fig. 9.3).

9.3   Kinetic energy

Definition 9.5 (Kinetic energy). For a particle with mass m and velocity $\bar{v}$ the *kinetic energy* is

$$K \equiv \frac{1}{2}m(\bar{v} \cdot \bar{v}) = \frac{1}{2}mv^2. \quad (9.6)$$

Forces acting on a particle may change its velocity, and thus change its kinetic energy. How the work of a force is transformed into kinetic energy is described in the *Work–energy theorem*.

Theorem 9.6 (Work–energy theorem). For a particle with mass m, subjected to a force sum $\Sigma \bar{F}$ between positions 1 and 2 (Fig. 9.4), it holds that

$$\Sigma U_{1-2} = K_2 - K_1, \quad (9.7)$$

where ΣU_{1-2} is the work of $\Sigma \bar{F}$, and K_1 and K_2 denote the kinetic energy of positions 1 and 2, respectively.

Proof. Differentiation of the kinetic energy, Eq. (9.6), w.r.t. time gives

$$\begin{aligned}
\frac{\mathrm{d}K}{\mathrm{d}t} &= \frac{\mathrm{d}}{\mathrm{d}t}\left(\frac{1}{2}m\bar{v} \cdot \bar{v}\right) = \{\text{Product rule (A.25b)}\} \\
&= \frac{1}{2}m\bar{a} \cdot \bar{v} + \frac{1}{2}m\bar{v} \cdot \bar{a} \\
&= m\bar{a} \cdot \bar{v} = \{\text{Law of force and acceleration}\} \\
&= \Sigma \bar{F} \cdot \bar{v}.
\end{aligned} \quad (9.8)$$

Using Eq. (A.30), this relation can be expressed as

$$\Sigma \bar{F} \cdot \bar{v} \mathrm{d}t = dK \quad \Leftrightarrow \quad \{\text{Theorem A.3}\} \quad \Leftrightarrow$$

$$\int_{t_1}^{t_2} \Sigma \bar{F} \cdot \bar{v} \mathrm{d}t = \int_{K_1}^{K_2} dK \quad \Leftrightarrow \quad \{\text{Def. 9.4}\} \quad \Leftrightarrow$$

$$\Sigma U_{1-2} = K_2 - K_1,$$

where t_1 and t_2 denote the times of positions 1 and 2, respectively. $\square$

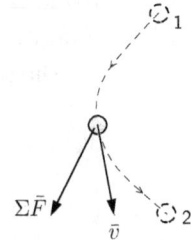

Figure 9.3: The direction of the motion of a particle relative to a space-fixed obstruction is perpendicular to the normal force $\bar{N}$.

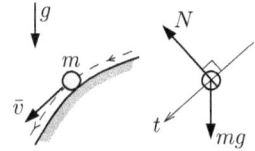

Figure 9.4: Geometry for the Work–energy theorem for the path of a particle between positions 1 and 2.

9.4 Conservative forces

Conservative forces preserve the total *mechanical energy* when they carry out work on a particle. The mechanical energy is the sum of the kinetic energy, the potential energy and the elastic energy.

The work of conservative forces does not transform mechanical energy into other forms of energy, such as heat or electromagnetic energy. Friction generates heat, and is therefore not a conservative force. In contrast, the force of gravity is conservative.

Definition 9.7 (Potential energy of gravity). The *potential energy* of a particle with mass m in a uniform field of gravity $\bar{g} = -g\bar{e}_y$ is

$$V_{\mathrm{g}}(y) \equiv mgy, \tag{9.9}$$

where y is a *height coordinate*[16] relative to an inertial system.

[16] Also called *altitude coordinate*.

Thus, for a terrestrial coordinate system, the potential energy of a particle increases linearly with distance above an arbitrary ground level.

Theorem 9.8 (Work of the force of gravity). For a particle $\mathcal{P}$ with mass m in a uniform field of gravity $\bar{g} = -g\bar{e}_y$, the work of the force of gravity $m\bar{g}$ on the particle, between positions 1 and 2, is (Fig. 9.5)

$$U_{1-2} = -\left[V_{\mathrm{g}}(y_2) - V_{\mathrm{g}}(y_1)\right], \tag{9.10}$$

where y_1 and y_2 are the height coordinates of positions 1 and 2, respectively, and $V_{\mathrm{g}}(y)$ is the potential energy of the particle.

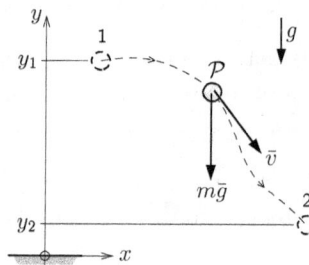

Figure 9.5: Geometry for the work of the force of gravity on a particle $\mathcal{P}$.

Proof. Let t_1 and t_2 be the times corresponding to positions 1 and 2. Then, according to Def. 9.2, we have

$$
\begin{aligned}
U_{1-2} &= \int_{t_1}^{t_2} m\bar{g} \cdot \bar{v}\,\mathrm{d}t = \left\{\text{Eq. (7.8)}\right\} \\
&= \int_{t_1}^{t_2} -mg\bar{e}_y \cdot (\dot{x}\bar{e}_x + \dot{y}\bar{e}_y + \dot{z}\bar{e}_z)\,\mathrm{d}t \\
&= -mg \int_{t_1}^{t_2} \frac{\mathrm{d}y}{\mathrm{d}t}\,\mathrm{d}t = \left\{ \begin{array}{l} \text{subst. (A.39)} \\ y = y(t),\ dy = \frac{\mathrm{d}y}{\mathrm{d}t}dt \end{array} \right\} \\
&= -mg \int_{y_1}^{y_2} \mathrm{d}y \\
&= -mg(y_2 - y_1) = \left\{\text{Def. 9.7}\right\} \\
&= -\left[V_{\mathrm{g}}(y_2) - V_{\mathrm{g}}(y_1)\right].
\end{aligned}
$$
$\square$

Springs, *e.g.* coil springs (Fig. 9.6), can be used to store mechanical energy. The spring force is conservative.

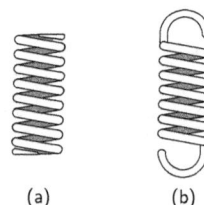

(a)   (b)

Figure 9.6: Coil springs designed for (a) compression, and (b) tension.

Definition 9.9 (Elastic energy of a linear spring). The *elastic energy* of a linear spring with spring constant k and natural length ℓ_0 (Fig. 1.5) is

$$V_e(\ell) \equiv \frac{1}{2}k(\ell - \ell_0)^2, \tag{9.11}$$

where ℓ denotes the length of the deformed spring.

Alternatively, if we let $\delta = \ell - \ell_0$ denote the elongation of the spring, the elastic energy can be written

$$V_e = \frac{1}{2}k\delta^2. \tag{9.12}$$

Theorem 9.10 (Work of the spring force). For a particle $\mathcal{P}$ connected by a linear spring to a space-fixed point $\mathcal{O}$, the work of the spring force $\bar{F}_e$ on the particle, between positions 1 and 2, is (fig. 9.7)

$$U_{1-2} = -\left[V_e(\ell_2) - V_e(\ell_1)\right], \tag{9.13}$$

where ℓ_1 and ℓ_2 are the respective lengths of the spring in positions 1 and 2, and $V_e(\ell)$ is the elastic energy of the spring.

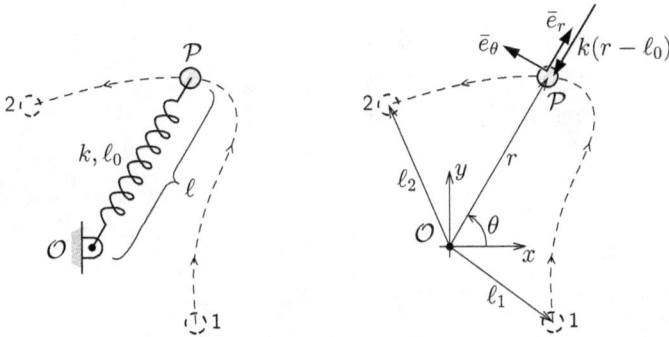

Figure 9.7: Geometry for the work of a spring force on a particle $\mathcal{P}$.

Proof. We introduce a polar coordinate system with origin $\mathcal{O}$. Then the spring force is $\bar{F}_e = -k(r - \ell_0)\bar{e}_r$, where k is the spring constant and ℓ_0 is the natural length of the spring. Let t_1 and t_2 denote the times corresponding to positions 1 and 2. Then, according to Def. 9.2, we have

$$U_{1-2} = \int_{t_1}^{t_2} \bar{F}_e \cdot \bar{v}\,\mathrm{d}t = \left\{\text{Eq. (7.17)}\right\}$$

$$= \int_{t_1}^{t_2} -k(r - \ell_0)\bar{e}_r \cdot (\dot{r}\bar{e}_r + r\dot{\theta}\bar{e}_\theta)\,\mathrm{d}t$$

$$= -\int_{t_1}^{t_2} k[r(t) - \ell_0]\frac{\mathrm{d}r}{\mathrm{d}t}\,\mathrm{d}t = \left\{\begin{array}{l}\text{subst. (A.39)} \\ u = r(t) - \ell_0,\ du = \frac{\mathrm{d}r}{\mathrm{d}t}dt\end{array}\right\}$$

$$= -\int_{r(t_1)-\ell_0}^{r(t_2)-\ell_0} ku\,\mathrm{d}u = \{r(t_1) = \ell_1,\, r(t_2) = \ell_2\}$$

$$= -\left[\frac{1}{2}ku^2\right]_{\ell_1-\ell_0}^{\ell_2-\ell_0}$$

$$= -\left[\frac{1}{2}k(\ell_2 - \ell_0)^2 - \frac{1}{2}k(\ell_1 - \ell_0)^2\right] = \{\text{Def. } 9.9\}$$

$$= -\left[V_{\mathrm{e}}(\ell_2) - V_{\mathrm{e}}(\ell_1)\right]. \qquad\qquad \square$$

9.5   Work–energy theorem with potentials

The work of the force of gravity or a spring force can be calculated using their corresponding potentials, V_{g} and V_{e}. The work of other forces must be integrated using Eq. (9.3). For this reason, the Work–energy theorem can be reformulated as follows.

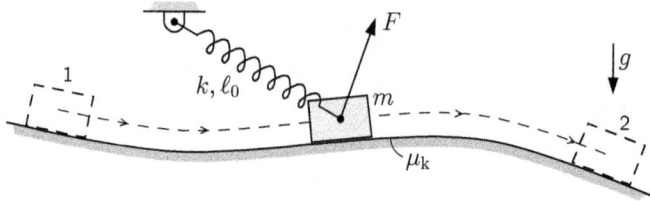

Figure 9.8: A particle affected by gravity, a spring force, and remaining forces $\Sigma\bar{F}'$ moves along a given path. The remaining forces include, for instance, the friction force and the external force $\bar{F}$.

Consider a particle with mass m and velocity $\bar{v}$, which is affected by the force of gravity $m\bar{g}$, a spring force $\bar{F}_{\mathrm{e}}$, and remaining forces $\Sigma\bar{F}'$ while moving between positions 1 and 2 (Fig. 9.8). According to Theorem 9.6, we have

$$\int_{t_1}^{t_2} (m\bar{g} + \bar{F}_{\mathrm{e}} + \Sigma\bar{F}') \cdot \bar{v}\,\mathrm{d}t = K_2 - K_1 \quad\Leftrightarrow$$

$$\int_{t_1}^{t_2} m\bar{g} \cdot \bar{v}\,\mathrm{d}t + \int_{t_1}^{t_2} \bar{F}_{\mathrm{e}} \cdot \bar{v}\,\mathrm{d}t + \int_{t_1}^{t_2} \Sigma\bar{F}' \cdot \bar{v}\,\mathrm{d}t = K_2 - K_1 \quad\Leftrightarrow$$

$$-(V_{\mathrm{g}2} - V_{\mathrm{g}1}) - (V_{\mathrm{e}2} - V_{\mathrm{e}1}) + \int_{t_1}^{t_2} \Sigma\bar{F}' \cdot \bar{v}\,\mathrm{d}t = K_2 - K_1, \qquad (9.14)$$

where we used Theorems 9.8 and 9.10. If we let

$$\Sigma U'_{1-2} = \int_{t_1}^{t_2} \Sigma\bar{F}' \cdot \bar{v}\,\mathrm{d}t$$

denote the work done by all forces, excluding forces of gravity and spring forces, then Eq. (9.14) can be written as

$$\Sigma U'_{1-2} = (V_{\mathrm{g}2} - V_{\mathrm{g}1}) + (V_{\mathrm{e}2} - V_{\mathrm{e}1}) + (K_2 - K_1). \qquad (9.15)$$

In problem solving, one calculates the left-hand side of Eq. (9.15) using Eq. (9.3), while the right-hand side is calculated using the definitions for the potential energy, the elastic energy, and the kinetic energy.

10

Momentum and angular momentum of particles

It is sometimes difficult to express the force on a particle as a function of its position. In such situations, we need an alternative to the Work–energy method for the analysis. For this purpose, we introduce the quantities of *momentum* and *angular momentum* of particles.

10.1 Momentum and impulse

Definition 10.1 (Momentum). The *momentum* of a particle with mass m and velocity $\bar{v}$ is (Fig. 10.1)

$$\bar{G} \equiv m\bar{v}. \tag{10.1}$$

The units of momentum are derived as

$$1\,\text{N·s} = 1\,\frac{\text{kg·m}}{\text{s}}, \qquad 1\,\text{lb}_{\text{f}}\text{·s} \approx 4.4482\,\text{N·s}.$$

In the following, we always assume that the particle mass m is constant. For such a particle, it holds that $\mathrm{d}\bar{G}/\mathrm{d}t = m\bar{a}$, so that the Law of force and acceleration can be written

$$\Sigma\bar{F} = \frac{\mathrm{d}\bar{G}}{\mathrm{d}t}. \tag{10.2}$$

Thus, if a force sum acts on a particle during a time interval, then the particle will change its momentum.

Theorem 10.2 (Impulse–momentum relation). If a particle is affected by a force sum $\Sigma\bar{F}$ between times t_1 and t_2, then it holds that

$$\int_{t_1}^{t_2} \Sigma\bar{F}\mathrm{d}t = \bar{G}(t_2) - \bar{G}(t_1), \tag{10.3}$$

where $\bar{G}$ is the momentum of the particle.

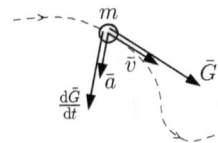

Figure 10.1: The direction of the momentum coincides with the velocity of the particle, while the time derivative of the momentum coincides with the acceleration.

Proof. Starting from the Law of force and acceleration, Eq. (10.2), we have

$$\Sigma \bar{F} = \frac{d\bar{G}}{dt} \quad \Leftrightarrow \quad \{Eq. (A.35)\} \quad \Leftrightarrow$$

$$\Sigma \bar{F} dt = d\bar{G} \quad \Leftrightarrow \quad \{Eq. (A.36)\} \quad \Leftrightarrow$$

$$\int_{t_1}^{t_2} \Sigma \bar{F} dt = \bar{G}(t_2) - \bar{G}(t_1). \qquad \qquad \square$$

The time integral on the left-hand side of the Impulse–momentum relation, Eq. (10.3), is called the *impulse* of the force sum.

Definition 10.3 (Impulse of a force). A force $\bar{F}$ with point of application $\mathcal{P}$, acting between times t_1 and t_2, yields an *impulse*

$$\bar{L} \equiv \int_{t_1}^{t_2} \bar{F} dt, \qquad \qquad (10.4)$$

which has the same point of application $\mathcal{P}$ as the force.

If several forces $\bar{F}_i$, $i = 1, \ldots, n$, contribute to the force sum that acts on a particle during the time interval $t_1 \le t \le t_2$, then each force gives an impulse $\bar{L}_i = \int_{t_1}^{t_2} \bar{F}_i dt$. Thus, the Impulse–momentum relation, Eq. (10.3), can be written as

$$\int_{t_1}^{t_2} \Sigma \bar{F} dt = \sum_{i=1}^{n} \int_{t_1}^{t_2} \bar{F}_i dt = \sum_{i=1}^{n} \bar{L}_i = \bar{G}(t_2) - \bar{G}(t_1). \qquad (10.5)$$

10.2   Angular momentum

Definition 10.4 (Angular momentum). For a particle $\mathcal{P}$ with mass m and velocity $\bar{v}$, the *angular momentum* w.r.t. an arbitrary point $\mathcal{A}$ is

$$\bar{H}_{\mathcal{A}} \equiv \overline{\mathcal{A}\mathcal{P}} \times m\bar{v}. \qquad \qquad (10.6)$$

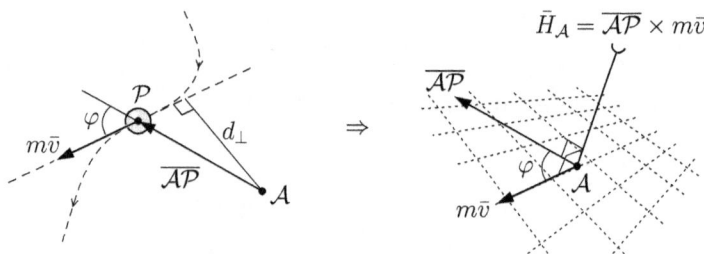

Figure 10.2: A particle with momentum $\bar{G} = m\bar{v}$, and angular momentum $\bar{H}_{\mathcal{A}}$ w.r.t. $\mathcal{A}$.

The direction of $\bar{H}_{\mathcal{A}}$ is given by the right-hand rule, and the magnitude of $\bar{H}_{\mathcal{A}}$ is

$$|\bar{H}_{\mathcal{A}}| = |\overline{\mathcal{A}\mathcal{P}} \times m\bar{v}| = |\overline{\mathcal{A}\mathcal{P}}||m\bar{v}| \sin \varphi = mvd_{\perp}, \qquad (10.7)$$

where φ is the angle between $\overline{\mathcal{AP}}$ and $\bar{v}$, and $d_\perp = |\overline{\mathcal{AP}}| \sin \varphi$ is the perpendicular distance from $\mathcal{A}$ to the line defined by the position and velocity of the particle (Fig. 10.2).

Theorem 10.5 (Moment–angular momentum relation). For a particle $\mathcal{P}$, affected by the force sum $\Sigma \bar{F}$, we have

$$\Sigma \bar{M}_\mathcal{D} = \frac{\mathrm{d}\bar{H}_\mathcal{D}}{\mathrm{d}t}, \tag{10.8}$$

where $\mathcal{D}$ is a space-fixed point, $\Sigma \bar{M}_\mathcal{D} = \overline{\mathcal{DP}} \times \Sigma \bar{F}$ is the moment sum w.r.t. $\mathcal{D}$, and $\bar{H}_\mathcal{D}$ is the angular momentum w.r.t. $\mathcal{D}$.

Proof. We choose a coordinate system with its origin in $\mathcal{D}$, so that $\bar{r} = \overline{\mathcal{DP}}$. According to Def. 10.4, with m the mass of the particle, we obtain

$$\begin{aligned}
\frac{\mathrm{d}\bar{H}_\mathcal{D}}{\mathrm{d}t} &= \frac{\mathrm{d}}{\mathrm{d}t} (\bar{r} \times m\bar{v}) = \left\{ \text{Product rule (A.25c)} \right\} \\
&= \frac{\mathrm{d}\bar{r}}{\mathrm{d}t} \times m\bar{v} + \bar{r} \times m\frac{\mathrm{d}\bar{v}}{\mathrm{d}t} = \left\{ \text{Defs. 7.2 and 7.3} \right\} \\
&= \bar{v} \times m\bar{v} + \bar{r} \times m\bar{a} \\
&= \bar{r} \times m\bar{a} = \left\{ \text{Law of force and acceleration} \right\} \\
&= \overline{\mathcal{DP}} \times \Sigma \bar{F} \\
&= \Sigma \bar{M}_\mathcal{D}. \qquad \qquad \qquad \qquad \qquad \square
\end{aligned}$$

If we integrate the Moment–angular momentum relation over time, we obtain a relation describing the change of angular momentum of a particle across a time interval:

Theorem 10.6 (Angular impulse–angular momentum relation). For a particle $\mathcal{P}$ affected by a force sum $\Sigma \bar{F}$ between times t_1 and t_2, we have

$$\int_{t_1}^{t_2} \Sigma \bar{M}_\mathcal{D}\mathrm{d}t = \bar{H}_\mathcal{D}(t_2) - \bar{H}_\mathcal{D}(t_1), \tag{10.9}$$

where $\mathcal{D}$ is a space-fixed point, $\Sigma \bar{M}_\mathcal{D} = \overline{\mathcal{DP}} \times \Sigma \bar{F}$ is the moment sum w.r.t. $\mathcal{D}$, and $\bar{H}_\mathcal{D}$ is the angular momentum of the particle w.r.t. $\mathcal{D}$.

Proof. According to Eq. (10.8), we have

$$\begin{aligned}
\Sigma \bar{M}_\mathcal{D} = \frac{\mathrm{d}\bar{H}_\mathcal{D}}{\mathrm{d}t} \quad &\Leftrightarrow \quad \left\{ \text{Eq. (A.35)} \right\} \quad \Leftrightarrow \\
\Sigma \bar{M}_\mathcal{D}dt = d\bar{H}_\mathcal{D} \quad &\Leftrightarrow \quad \left\{ \text{Eq. (A.36)} \right\} \quad \Leftrightarrow \\
\int_{t_1}^{t_2} \Sigma \bar{M}_\mathcal{D}dt = \bar{H}_\mathcal{D}(t_2) - \bar{H}_\mathcal{D}(t_1)&. \qquad \qquad \qquad \square
\end{aligned}$$

The integral on the left-hand side of Eq. (10.9) is the *angular impulse* of the particle w.r.t. $\mathcal{D}$ between t_1 and t_2.

Angular momentum for planar motion

Consider a particle $\mathcal{P}$ with mass m, that moves in a reference plane with normal $\bar{e}_{\mathrm{n}}$. Thus, the velocity vector $\bar{v}$ of this particle $\mathcal{P}$ lies in the reference plane. According to Def. 10.4, the angular momentum of the particle, w.r.t. a point $\mathcal{A}$ in the reference plane, is $\bar{H}_{\mathcal{A}} = \overline{\mathcal{AP}} \times m\bar{v}$. Since both $\overline{\mathcal{AP}}$ and $\bar{v}$ lie in the reference plane, it holds that $\bar{H}_{\mathcal{A}} = H_{\mathcal{A}}\bar{e}_{\mathrm{n}}$ with

$$H_{\mathcal{A}} = \pm|\overline{\mathcal{AP}} \times m\bar{v}| = \{\text{Eq. (A.19)}\}$$
$$= \pm m|\overline{\mathcal{AP}}||\bar{v}|\sin\varphi,$$

where φ is the angle between $\overline{\mathcal{AP}}$ and $\bar{v}$ (Fig. 10.3). We let $d_{\perp} = |\overline{\mathcal{AP}}|\sin\varphi$ be the perpendicular distance from $\mathcal{A}$ to the line defined by $\mathcal{P}$ and the velocity vector. It follows that

$$H_{\mathcal{A}} = \pm mvd_{\perp}. \tag{10.10}$$

The direction of the angular momentum is given by the right-hand rule (*cf.* moment of force in Sect. 2.4). The counterclockwise oriented angular momentum $H_{\mathcal{A}}$, illustrated in Fig. 10.3, is oriented in the $\bar{e}_z$ direction. If we choose the normal of the reference plane to be $\bar{e}_{\mathrm{n}} = \bar{e}_z$, then the angular momentum $H_{\mathcal{A}}$ will have a positive sign in its scalar representation. Conversely, a clockwise oriented angular momentum will have a negative sign.

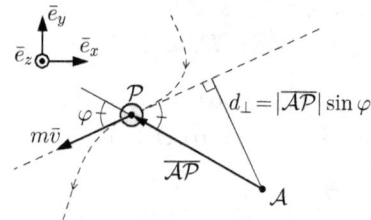

Figure 10.3: The geometry of the angular momentum for planar motion, where the xy plane is the reference plane.

10.3   Systems of particles

A *system of particles* contains several particles with different masses and different paths of motion:

Definition 10.7 (System of particles). A *system of particles*, is a set of particles $\mathcal{P}_i$, $i = 1, \ldots, n$, with masses m_i, position vectors $\bar{r}_i$ and velocities $\bar{v}_i$ (Fig. 10.4).

Definition 10.8 (Momentum of a system of particles). A system of particles, with notation as in Def. 10.7, has the momentum

$$\Sigma\bar{G} \equiv \sum_{i=1}^{n} m_i\bar{v}_i. \tag{10.11}$$

Definition 10.9 (Angular momentum of a system of particles). For a system of particles, with notation as in Def. 10.7, the angular momentum w.r.t. an arbitrary point $\mathcal{A}$ is

$$\Sigma\bar{H}_{\mathcal{A}} \equiv \sum_{i=1}^{n} \overline{\mathcal{AP}}_i \times m_i\bar{v}_i. \tag{10.12}$$

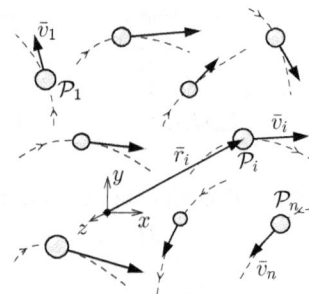

Figure 10.4: A system of n different particles $\mathcal{P}_i$, $i = 1, \ldots, n$.

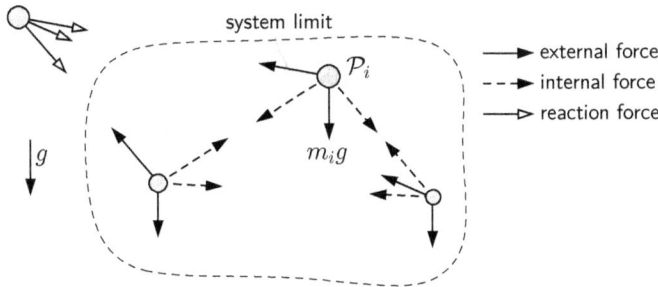

Figure 10.5: Pairwise interaction between particles in a system of particles creates internal forces. Interaction with objects outside the system creates external forces.

Consider a system of particles, $\mathcal{P}_1, \ldots, \mathcal{P}_n$. Pairwise interaction between the particles within this system gives rise to *internal forces*. Moreover, interaction with objects outside the system of particles gives rise to *external forces* (Fig. 10.5), including forces of gravity.[17] Let the sum of internal forces acting on $\mathcal{P}_i$ be denoted by $\bar{F}_i^{\text{int}}$, and let the sum of external forces acting on $\mathcal{P}_i$ be denoted by $\bar{F}_i^{\text{ext}}$ (Fig. 10.6). According to the Law of action and reaction, and to Postulate 8.4, each pairwise interaction within the system of particles creates a force pair with a zero couple. Thus, the internal forces form a zero system,

[17] Gravitation is external when the Earth is placed outside the system.

$$\sum_{i=1}^{n} \bar{F}_i^{\text{int}} = \bar{0}, \tag{10.13a}$$

$$\sum_{i=1}^{n} \overline{\mathcal{AP}}_i \times \bar{F}_i^{\text{int}} = \bar{0}, \tag{10.13b}$$

for every reference point $\mathcal{A}$.

Theorem 10.10 (Impulse–momentum relation for a system of particles). For a system of particles, affected by the sum $\Sigma \bar{F}^{\text{ext}}$ of external forces between times t_1 and t_2, it holds that

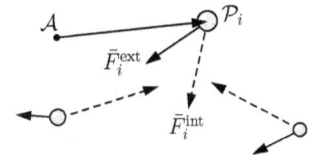

Figure 10.6: A particle $\mathcal{P}_i$ affected by an internal force sum $\bar{F}_i^{\text{int}}$, and an external force sum $\bar{F}_i^{\text{ext}}$, *cf.* Fig. 10.5.

$$\int_{t_1}^{t_2} \Sigma \bar{F}^{\text{ext}} \mathrm{d}t = \Sigma \bar{G}(t_2) - \Sigma \bar{G}(t_1), \tag{10.14}$$

where $\Sigma \bar{G}$ is the momentum of the system of particles.

Proof. Let the internal and external force sums that act on particle $\mathcal{P}_i$ be denoted by $\bar{F}_i^{\text{int}}$ and $\bar{F}_i^{\text{ext}}$, respectively (Fig. 10.6). The Impulse–momentum relation, Eq. (10.3), then gives

$$\int_{t_1}^{t_2} \left(\bar{F}_i^{\text{int}} + \bar{F}_i^{\text{ext}} \right) \mathrm{d}t = \bar{G}_i(t_2) - \bar{G}_i(t_1), \qquad i = 1, \ldots, n.$$

Summation over i gives

$$\sum_{i=1}^{n} \int_{t_1}^{t_2} \left(\bar{F}_i^{\text{int}} + \bar{F}_i^{\text{ext}}\right) \mathrm{d}t = \sum_{i=1}^{n} \bar{G}_i(t_2) - \sum_{i=1}^{n} \bar{G}_i(t_1) \quad \Leftrightarrow \quad \{\text{Def. 10.8}\} \quad \Leftrightarrow$$

$$\int_{t_1}^{t_2} \left(\underbrace{\sum_{i=1}^{n} \bar{F}_i^{\text{int}}}_{=\bar{0}} + \underbrace{\sum_{i=1}^{n} \bar{F}_i^{\text{ext}}}_{=\Sigma \bar{F}^{\text{ext}}}\right) \mathrm{d}t = \Sigma \bar{G}(t_2) - \Sigma \bar{G}(t_1) \quad \Leftrightarrow$$

$$\int_{t_1}^{t_2} \Sigma \bar{F}^{\text{ext}} \mathrm{d}t = \Sigma \bar{G}(t_2) - \Sigma \bar{G}(t_1),$$

where the sum of internal forces is $\bar{0}$, since the internal forces form a zero system. $\square$

Theorem 10.10 is valid even when the particles collide against each other, so that heat develops and mechanical energy is lost. A special case occurs for a system of particles that is not affected by any external forces, $\Sigma \bar{F}^{\text{ext}} = \bar{0}$. In such a case, the momentum of the system of particles is conserved:

$$\Sigma \bar{G}(t_2) = \Sigma \bar{G}(t_1). \tag{10.15}$$

Theorem 10.11 (Angular impulse–angular momentum relation for a system of particles). For a system of particles affected by a sum $\Sigma \bar{M}_{\mathcal{D}}^{\text{ext}}$ of external moments w.r.t. a space-fixed point $\mathcal{D}$ between times t_1 and t_2, it holds that

$$\int_{t_1}^{t_2} \Sigma \bar{M}_{\mathcal{D}}^{\text{ext}} \mathrm{d}t = \Sigma \bar{H}_{\mathcal{D}}(t_2) - \Sigma \bar{H}_{\mathcal{D}}(t_1), \tag{10.16}$$

where $\Sigma \bar{H}_{\mathcal{D}}$ is the angular momentum of the system of particles w.r.t. $\mathcal{D}$.

Proof. Let the internal and the external force sums acting on particle $\mathcal{P}_i$ be denoted by $\bar{F}_i^{\text{int}}$ and $\bar{F}_i^{\text{ext}}$, respectively (Fig. 10.6). The Angular impulse–angular momentum relation, Eq. (10.9), then gives

$$\int_{t_1}^{t_2} \overline{\mathcal{D}\mathcal{P}}_i \times (\bar{F}_i^{\text{int}} + \bar{F}_i^{\text{ext}}) \mathrm{d}t = \bar{H}_{\mathcal{D}i}(t_2) - \bar{H}_{\mathcal{D}i}(t_1), \qquad i = 1, \ldots, n,$$

where $\bar{H}_{\mathcal{D}i}$ denotes the angular momentum of particle $\mathcal{P}_i$ w.r.t. $\mathcal{D}$. Summation over i gives

$$\sum_{i=1}^{n} \int_{t_1}^{t_2} \left(\overline{\mathcal{D}\mathcal{P}}_i \times \bar{F}_i^{\text{int}} + \overline{\mathcal{D}\mathcal{P}}_i \times \bar{F}_i^{\text{ext}}\right) \mathrm{d}t = \sum_{i=1}^{n} \bar{H}_{\mathcal{D}i}(t_2) - \sum_{i=1}^{n} \bar{H}_{\mathcal{D}i}(t_1) \quad \Leftrightarrow \quad \{\text{Def. 10.9}\} \quad \Leftrightarrow$$

$$\int_{t_1}^{t_2} \left(\underbrace{\sum_{i=1}^{n} \overline{\mathcal{D}\mathcal{P}}_i \times \bar{F}_i^{\text{int}}}_{=\bar{0}} + \underbrace{\sum_{i=1}^{n} \overline{\mathcal{D}\mathcal{P}}_i \times \bar{F}_i^{\text{ext}}}_{=\Sigma \bar{M}_{\mathcal{D}}^{\text{ext}}}\right) \mathrm{d}t = \Sigma \bar{H}_{\mathcal{D}}(t_2) - \Sigma \bar{H}_{\mathcal{D}}(t_1) \quad \Leftrightarrow$$

$$\int_{t_1}^{t_2} \Sigma \bar{M}_{\mathcal{D}}^{\text{ext}} \mathrm{d}t = \Sigma \bar{H}_{\mathcal{D}}(t_2) - \Sigma \bar{H}_{\mathcal{D}}(t_1),$$

where the sum of internal moments is $\bar{0}$, since the internal forces form a zero system. □

An important special case of Theorem 10.11 occurs when a space-fixed point $\mathcal{D}$ such that $\Sigma \bar{M}_{\mathcal{D}}^{\text{ext}} = \bar{0}$ can be identified (Fig. 10.7). In such a case, the angular momentum w.r.t. $\mathcal{D}$ is conserved for the system of particles:

$$\Sigma \bar{H}_{\mathcal{D}}(t_2) = \Sigma \bar{H}_{\mathcal{D}}(t_1). \tag{10.17}$$

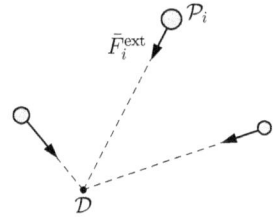

Figure 10.7: An example where the sum of the external moments of force on a system of particles is zero w.r.t. a space-fixed point $\mathcal{D}$.

10.4 Impact

Collisions between particles, and collisions between particles and space-fixed objects, are examples of impacts. Despite the complication of heat production, it is under certain conditions possible to predict the kinetics of the impact process.

Conservation of momentum

Consider two particles, $\mathcal{P}$ and $\mathcal{Q}$, colliding with each other. During this event, their velocities will experience great change during a relatively short time interval. This type of process is called an *impact*.

The impact between $\mathcal{P}$ and $\mathcal{Q}$ is assumed to occur in a time interval $0 \leq t \leq \Delta t$. Let $\bar{v}_{\mathcal{P}}$ and $\bar{v}_{\mathcal{Q}}$ be the velocities of $\mathcal{P}$ and $\mathcal{Q}$ at time $t = 0$, before the impact. Furthermore, let $\bar{v}'_{\mathcal{P}}$ and $\bar{v}'_{\mathcal{Q}}$ be the velocities at time $t = \Delta t$ after the impact (Fig. 10.8). Henceforth, we will use the prime symbol to denote quantities evaluated after the impact.

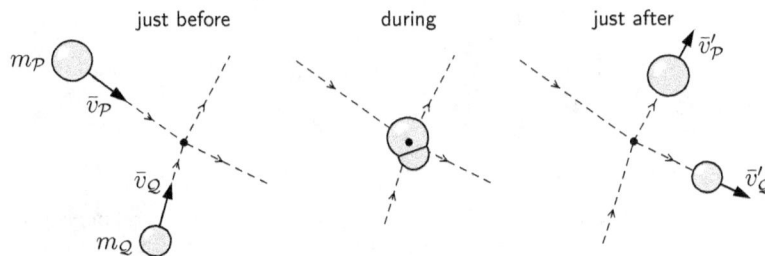

Figure 10.8: Impact between two particles, $\mathcal{P}$ and $\mathcal{Q}$.

Considering Eq. (10.11), the momentum of the system of particles before and after the impact can be written

$$\Sigma \bar{G} = m_{\mathcal{P}} \bar{v}_{\mathcal{P}} + m_{\mathcal{Q}} \bar{v}_{\mathcal{Q}},$$
$$\Sigma \bar{G}' = m_{\mathcal{P}} \bar{v}'_{\mathcal{P}} + m_{\mathcal{Q}} \bar{v}'_{\mathcal{Q}},$$

respectively, where $m_{\mathcal{P}}$ and $m_{\mathcal{Q}}$ are the masses of the particles. According to the Impulse–momentum relation for a system of particles,

Theorem 10.10, it holds that

$$\int_0^{\Delta t} \Sigma \bar{F}^{\text{ext}} \mathrm{d}t = \Sigma \bar{G}' - \Sigma \bar{G},$$

where $\Sigma \bar{F}^{\text{ext}}$ is the sum of external forces acting on the particles.

In an *instantaneous impact model* of a two-particle system, one assumes that the duration Δt of the impact is sufficiently small, so that the external impulse on the system of particles, including the forces of gravity, can be neglected:

$$\int_0^{\Delta t} \Sigma \bar{F}^{\text{ext}} \mathrm{d}t = \bar{0}.$$

Whether such an approximation is reasonable needs to be assessed from case to case. One consequence of the instantaneous impact model is that the momentum of the system of particles is conserved during impact:

$$\Sigma \bar{G}' = \Sigma \bar{G}. \tag{10.18}$$

Direct central impact

During a *direct central impact*, two particles $\mathcal{P}$ and $\mathcal{Q}$ travel along a common straight line, both before and after the impact. We introduce a coordinate x along this line of motion, and let $v_{\mathcal{P}}$ and $v_{\mathcal{Q}}$ denote the signed velocities of the particles before the impact, while $v'_{\mathcal{P}}$ and $v'_{\mathcal{Q}}$ denote the velocities after the impact (Fig. 10.9).

We employ an instantaneous impact model, so that the momentum is conserved during impact, $\Sigma G'_x = \Sigma G_x$:

$$\rightarrow^x: \quad m_{\mathcal{P}} v'_{\mathcal{P}} + m_{\mathcal{Q}} v'_{\mathcal{Q}} = m_{\mathcal{P}} v_{\mathcal{P}} + m_{\mathcal{Q}} v_{\mathcal{Q}}. \tag{10.19}$$

This model neglects the impulses of external forces, *e.g.* the force of gravity, on the particles.

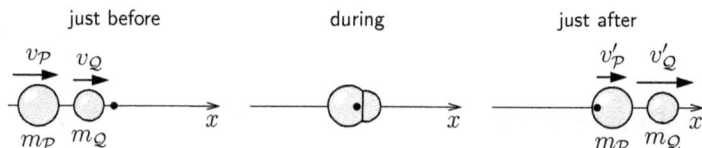

Figure 10.9: Direct central impact, where two particles, $\mathcal{P}$ and $\mathcal{Q}$, move along a straight line and collide.

Even though the masses and velocities of the particles are known before the direct central impact, it is not possible to calculate the velocities of the particles after the impact using only Eq. (10.19); one equation is not sufficient to determine the two unknowns, $v'_{\mathcal{P}}$ and $v'_{\mathcal{Q}}$. We need an additional relation to calculate the outcome of a direct central impact.

Empirical relation 10.12 (Coefficient of restitution). For direct central impact between two particles, $\mathcal{P}$ and $\mathcal{Q}$, whose velocities before the impact are $v_{\mathcal{P}}$ and $v_{\mathcal{Q}}$, and after the impact are $v'_{\mathcal{P}}$ are $v'_{\mathcal{Q}}$, it holds that

$$e = -\frac{v'_{\mathcal{Q}} - v'_{\mathcal{P}}}{v_{\mathcal{Q}} - v_{\mathcal{P}}}, \qquad (10.20)$$

where the constant $e \in [0,1]$ is the *coefficient of restitution.*

If the energy of the two-particle system is conserved during the impact, then the impact is *elastic*, and $e = 1$. If $e = 0$, then the impact is said to be *plastic*. If the coefficient of restitution is known, then Eqs. (10.19) and (10.20) can be solved for $v'_{\mathcal{P}}$ and $v'_{\mathcal{Q}}$.

Oblique impact

Consider two bodies that collide obliquely, with a nonzero angle between their paths of approach. In most cases, such an oblique impact induces rotation of the bodies, so that they should not be regarded as particles. However, special cases exist for which a particle model can be applied.

Consider two particles, $\mathcal{P}$ and $\mathcal{Q}$, that collide, so that the particles are joined together and continue along a common path (Fig. 10.10). Thus, the velocities of the particles after the impact become $\bar{v}'_{\mathcal{P}} = \bar{v}'_{\mathcal{Q}} = \bar{v}'$. We introduce an instantaneous impact model, for which the momentum is conserved during the impact, $\Sigma \bar{G}' = \Sigma \bar{G}$:

$$(m_{\mathcal{P}} + m_{\mathcal{Q}})\bar{v}' = (m_{\mathcal{P}}\bar{v}_{\mathcal{P}} + m_{\mathcal{Q}}\bar{v}_{\mathcal{Q}}), \qquad (10.21)$$

using the notation in Fig. 10.10. If we know the velocities, $\bar{v}_{\mathcal{P}}$ and $\bar{v}_{\mathcal{Q}}$, of the particles before the impact, then we can determine the common velocity $\bar{v}'$ after the impact from Eq. (10.21).

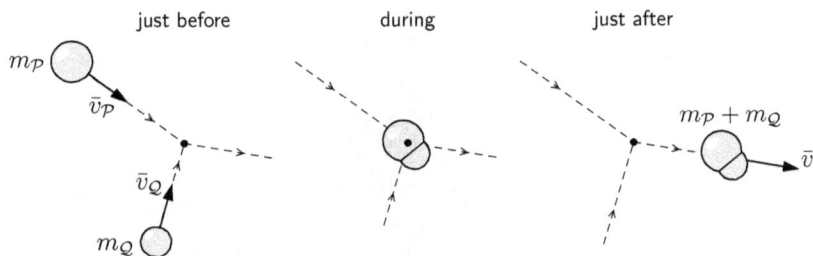

Figure 10.10: Oblique impact, where two particles, $\mathcal{P}$ and $\mathcal{Q}$, collide, so that the particles join together and continue along a common path.

Impact impulse

Consider a particle $\mathcal{P}$ colliding with a body, which is not necessarily a particle. Similarly to the case of colliding particles, the velocity of $\mathcal{P}$ will change considerably during a short time interval. This implies that the force sum acting on the particle grows very large during the impact.

Now, consider an impact occurring between times $t = 0$ and $t = \Delta t$. Let the momentum of $\mathcal{P}$ before the impact be $\bar{G} = m\bar{v}$, and after the impact be $\bar{G}' = m\bar{v}'$. The Impulse–momentum relation, Eq. (10.3), for a particle then gives

$$\underbrace{\int_0^{\Delta t} \bar{F}^{\mathrm{imp}}\mathrm{d}t}_{=\bar{L}^{\mathrm{imp}}} + \underbrace{\int_0^{\Delta t} \bar{F}_1\mathrm{d}t}_{=\bar{L}_1} + \cdots + \underbrace{\int_0^{\Delta t} \bar{F}_n\mathrm{d}t}_{=\bar{L}_n} = \bar{G}' - \bar{G}, \qquad (10.22)$$

where $\bar{F}^{\mathrm{imp}}$ denotes the impact force acting on $\mathcal{P}$ during the impact, while $\bar{F}_i$, $i = 1, \ldots, n$ are the other forces acting on $\mathcal{P}$.

We employ an instantaneous impact model (*cf.* Fig. 10.11). Since $\bar{F}^{\mathrm{imp}}$ grows large during the impact, its *impact impulse* $\bar{L}^{\mathrm{imp}} = \int_0^{\Delta t} \bar{F}^{\mathrm{imp}}\mathrm{d}t$ is nonnegligible. However, the duration Δt of the impact is sufficiently short so that all other impulses can be neglected, giving

$$\bar{L}^{\mathrm{imp}} = \bar{G}' - \bar{G}. \qquad (10.23)$$

Consequently, the impact impulse alone is assumed to create the sudden change in momentum of $\mathcal{P}$. This approximation is proper when $|\bar{L}^{\mathrm{imp}}| \gg |\bar{L}_i|$, $i = 1, \ldots, n$, meaning that the magnitude of the impact force is much greater than the magnitude of the other forces during the impact.

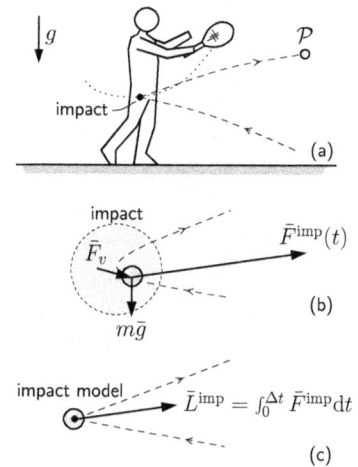

Figure 10.11: (a) Demonstration of the impact model. (b) Several forces, air resistance $\bar{F}_v$, the force of gravity $m\bar{g}$, and the impact force $\bar{F}^{\mathrm{imp}}$, act on a tennis ball during the strike. (c) Only the impact impulse $\bar{L}^{\mathrm{imp}}$, created by the impact force, is taken into account in an instantaneous impact model.

11
Harmonic oscillators

An oscillator is a mechanical system, that, when displaced from its equilibrium state, experiences a *restoring force* back towards its equilibrium. This restoring force could be due to any physical phenomenon, but herein it is represented by a linear spring.

11.1 Free harmonic oscillators

Undamped oscillator

Consider a carriage with mass m, rolling without friction on a horizontal surface. A linear spring with spring constant k connects this carriage to a wall (Fig. 11.1). The position of the carriage is described by a coordinate x, such that $x = 0$ when the spring is undeformed. In this case, since the coordinate is identical to the elongation of the spring, $x = \ell - \ell_0$, the spring force is $F_s = kx$. It is convenient to draw the free-body diagram for the special case $x > 0$. Such a diagram is automatically valid for $x \leq 0$ as well.

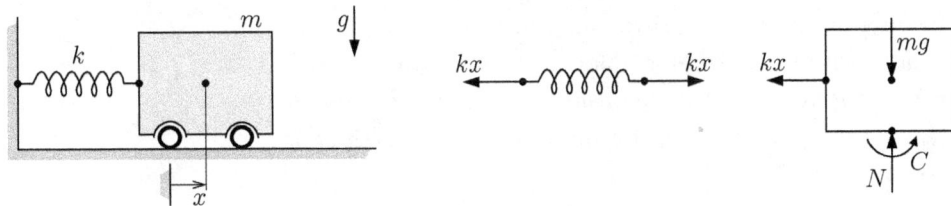

Figure 11.1: A carriage with mass m rolling horizontally without friction. A spring connects it to a wall. The carriage describes harmonic motion when it is brought out of its equilibrium position and released.

Since the motion is constrained to the x direction, the acceleration of the carriage is $\bar{a} = \ddot{x}\bar{e}_x$. Applying the Law of force and acceleration in

the x direction yields

$$\rightarrow^x: \qquad -kx = m\ddot{x} \quad \Leftrightarrow$$

$$\ddot{x} + \frac{k}{m}x = 0 \quad \Leftrightarrow$$

$$\ddot{x} + \omega_n^2 x = 0, \tag{11.1}$$

where ω_n is the *natural angular frequency*. In this example, we identify that $\omega_n = \sqrt{k/m}$.

The solution to the second-order, ordinary differential equation (11.1) is obtained by assuming that

$$x = x_0 + A\cos(\omega_n t) + B\sin(\omega_n t), \tag{11.2}$$

where x_0, A and B are real constants. We differentiate this assumed solution twice w.r.t. time and obtain

$$\dot{x} = -\omega_n A\sin(\omega_n t) + \omega_n B\cos(\omega_n t),$$

$$\ddot{x} = -\omega_n^2 A\cos(\omega_n t) - \omega_n^2 B\sin(\omega_n t).$$

By inserting x and $\ddot{x}$, we verify that Eq. (11.1) is satisfied for $x_0 = 0$, and for every choice of A and B. Therefore, Eq. (11.2) indeed describes the motion of the carriage. The constants A and B depend on the initial conditions of the system, that is the values of $x(0)$ and $\dot{x}(0)$ if the motion is initiated at $t = 0$.

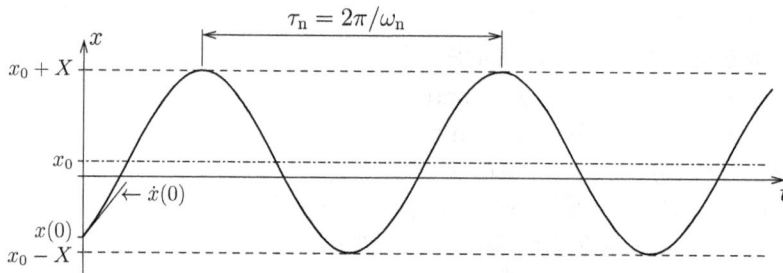

Figure 11.2: Example of undamped oscillation with initial conditions $x(0)$ and $\dot{x}(0)$. The amplitude of the motion is X, and the natural period is $\tau_n = 2\pi/\omega_n$.

The free, undamped oscillation described by Eq. (11.2) is called undamped harmonic motion. Its characteristics are illustrated in Fig. 11.2. The particle oscillates with the natural angular frequency ω_n and a constant amplitude X about an *equilibrium position* at which the force sum is zero. The *natural period*, that is the duration between two maxima of the harmonic motion, is given by[18]

$$\tau_n = \frac{2\pi}{\omega_n}. \tag{11.3}$$

The *amplitude* of the harmonic motion is half the difference between the maximum and minimum of $x(t)$. According to Eqs. (A.6) and (A.7),

[18] The *natural frequency* defined as

$$f_n = \frac{\omega_n}{2\pi} = \frac{1}{\tau_n},$$

with units of *hertz* (Hz) is also used in to quantify oscillations, where $1\,\text{Hz} = 1\,\text{s}^{-1}$.

the harmonic function in Eq. (11.2) can be written as

$$x(t) = x_0 + X\sin(\omega_n t + \psi), \qquad X = \sqrt{A^2 + B^2},$$

where X is the amplitude, for some value of ψ being the *phase angle*.

Damped harmonic oscillators

An ideal, free, undamped oscillation continues indefinitely with the same amplitude. However, in real, freely-oscillating systems, the amplitude eventually decreases due to *damping* caused by, for instance, friction or air resistance which transforms mechanical energy into heat. *Dampers* (Fig. 11.3a) are used to reduce the amplitude of oscillations or vibrations, that threaten the function of mechanical systems.

In Fig. 11.3b, a free-body diagram shows a linear damper with a damping force F_d acting in both ends. The damper has a current length ℓ, and a *damping coefficient* c with the units of N·s/m or lb$_f$·s/ft. The damping force of a linear damper is

$$F_d = c\dot{\ell}, \tag{11.4}$$

where $\dot{\ell}$ is the extension rate of the damper.

Again, consider a carriage with mass m, rolling without friction against a horizontal surface. A linear spring with spring constant k and a linear damper with damping coefficient c connect the carriage to a wall (Fig. 11.4). The position of the carriage is described by a coordinate x, such that $x = 0$ when the spring is undeformed. Since $\dot{x}$ is identical to the extension rate of the damper, the damping force is $F_d = c\dot{x}$. The force-couple system in the free-body diagram is drawn for the special case $x > 0$, $\dot{x} > 0$ (Fig. 11.4). This makes the diagram generally valid.

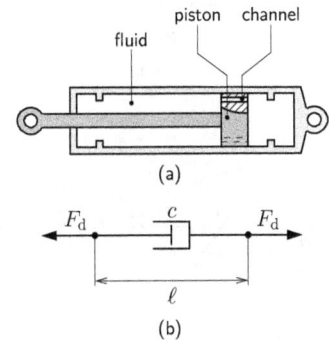

Figure 11.3: (a) A realization of a damper. The motion of the piston is restricted by liquid or gas that must pass through channels in the piston. (b) A free-body diagram for an ideal damper, where $F_d = c\dot{\ell}$.

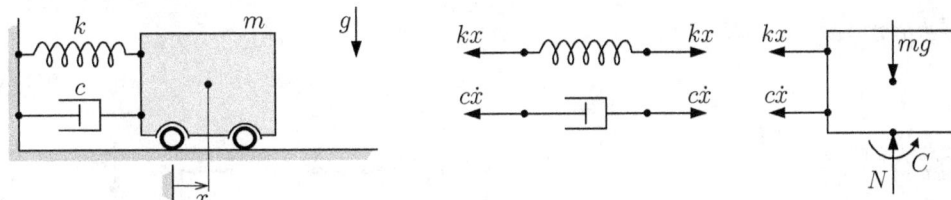

Figure 11.4: A carriage with mass m rolling horizontally without friction. It is connected by a spring and a damper to a wall. If it is disturbed from its equilibrium position, the carriage will describe damped harmonic oscillation.

Since the motion is constrained to the x direction, the acceleration of the carriage is $\bar{a} = \ddot{x}\bar{e}_x$, and the Law of force and acceleration gives

$$\rightarrow^x: \qquad -kx - c\dot{x} = m\ddot{x} \quad \Leftrightarrow$$

$$\ddot{x} + \frac{c}{m}\dot{x} + \frac{k}{m}x = 0 \quad \Leftrightarrow$$

$$\ddot{x} + 2\zeta\omega_n\dot{x} + \omega_n^2 x = 0, \tag{11.5}$$

where ω_n is the natural angular frequency, and $\zeta \geq 0$ is the *damping ratio*. In our example, we identify $\omega_n = \sqrt{k/m}$ and $\zeta = c/(2m\omega_n)$.

Equation (11.5) is a homogeneous,[19] second-order differential equation with constant coefficients . The form of its solution depends on the damping ratio:

$$
x(t) = \begin{cases}
Ae^{-\omega_n t(\zeta - \sqrt{\zeta^2 - 1})} + Be^{-\omega_n t(\zeta + \sqrt{\zeta^2 - 1})}, & \zeta > 1, \\
(A + Bt)e^{-\omega_n t}, & \zeta = 1, \\
[A\cos(\omega_d t) + B\sin(\omega_d t)]\,e^{-\zeta\omega_n t}, & \zeta < 1,
\end{cases} \tag{11.6}
$$

where $\omega_d = \omega_n\sqrt{1 - \zeta^2}$, and A and B are real constants. Consequently, there are three different types of damped oscillation called *overdamped*, $\zeta > 1$, *critically damped*, $\zeta = 1$, and *underdamped*, $\zeta < 1$ (Fig. 11.5).[20] Note that when $\zeta = 0$, the oscillator is underdamped, and that the expression for the undamped oscillation, *cf.* Eq. (11.2), is recovered.

[19] *Homogeneous differential equation –* all terms of the linear differential equation include $x(t)$ or its derivatives.

[20] The concepts of *strong damping*, $\zeta > 1$, and *weak damping*, $\zeta < 1$, are sometimes used.

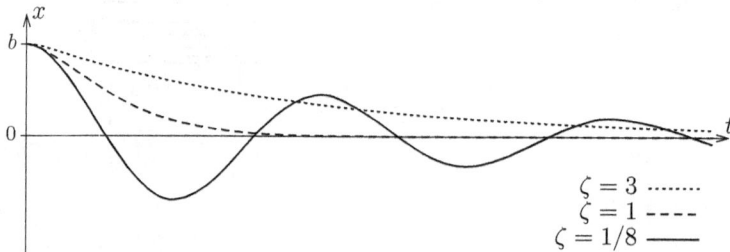

$\zeta = 3$ ········
$\zeta = 1$ - - - -
$\zeta = 1/8$ ——

Figure 11.5: An illustration of the motion of different free, damped oscillators, all with initial conditions $x(0) = b$ and $\dot{x}(0) = 0$: overdamped (dotted line), critically damped (dashed line), and underdamped (solid line).

Overdamped or critically damped oscillators will return to their equilibrium position without oscillation. This is obvious considering the forms of their solutions in Eq. (11.6), which do not include any harmonic factor. For underdamped systems, however, an oscillation that attenuates to the equilibrium position is observed.

11.2   Forced harmonic oscillators

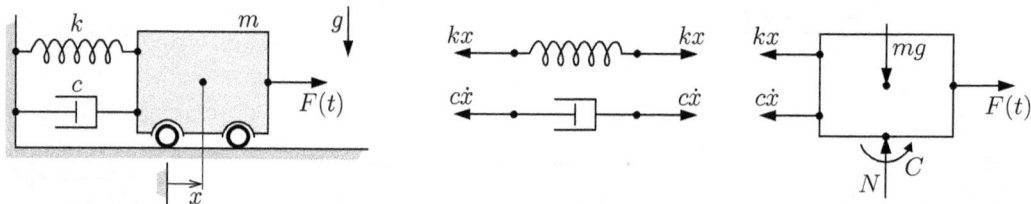

Figure 11.6: A carriage with mass m, connected to a wall by a spring and a damper. A force $F(t)$ keeps the carriage in motion.

Once again, consider a carriage with mass m, rolling without friction against a horizontal surface. The carriage is connected to a wall by a spring with spring constant k and a damper with damping coefficient c. Furthermore, a force $F(t)$ acts on the carriage (Fig. 11.6).

The Law of force and acceleration applied in the x direction yields

$$\rightarrow^x: \qquad -kx - c\dot{x} + F(t) = m\ddot{x} \quad \Leftrightarrow$$

$$\ddot{x} + \frac{c}{m}\dot{x} + \frac{k}{m}x = \frac{1}{m}F(t) \quad \Leftrightarrow$$

$$\ddot{x} + 2\zeta\omega_n\dot{x} + \omega_n^2 x = f(t), \tag{11.7}$$

where the right-hand side is a function $f(t) = F(t)/m$. Equation (11.7) is an inhomogeneous, second-order differential equation with constant coefficients. The general solution to this differential equation (11.7) can be written on the form

$$x(t) = x_h(t) + x_p(t), \tag{11.8}$$

where x_h is the *homogeneous solution*, and x_p is the *particular solution*.

The homogeneous solution x_h is the solution to

$$\ddot{x}_h + 2\zeta\omega_n\dot{x}_h + \omega_n^2 x_h = 0.$$

That is, the differential equation for x_h is identical to Eq. (11.5) for freely-damped oscillators. For this reason, the homogeneous solution is given by Eq. (11.6), as illustrated in Fig. 11.5 for different values of ζ. All homogeneous solutions attenuate to zero, since they, for every $\zeta > 0$, are dominated by a decreasing exponential factor:

$$x_h(t) \rightarrow 0, \quad \text{as} \quad t \rightarrow \infty.$$

Therefore, the forced, damped oscillation will, after a sufficiently long time, be described by its particular solution:

$$x(t) = x_h(t) + x_p(t) \rightarrow x_p(t), \quad \text{as} \quad t \rightarrow \infty.$$

Since the homogeneous solution attenuates while the particular solution persists, the particular solution is of interest in the study of forced harmonic oscillators.

Henceforth, we limit our discussion to cases where the applied force is a harmonic function, for instance, $F(t) = F_0 \sin(\omega t)$. Before we continue our analysis, we consider a numerical solution to Eq. (11.7) for the special case, $\zeta = 1/8$ and $\omega = \frac{5}{2}\omega_n$, with the initial conditions $x(0) = b$ and $\dot{x}(0) = 0$, as illustrated in Fig. 11.7. We observe an initial nonperiodic behavior called a *transient*. The motion later becomes periodic, corresponding to the particular solution, with an amplitude X and the angular frequency ω of the applied force. We wish to determine this amplitude X of the remaining oscillation.

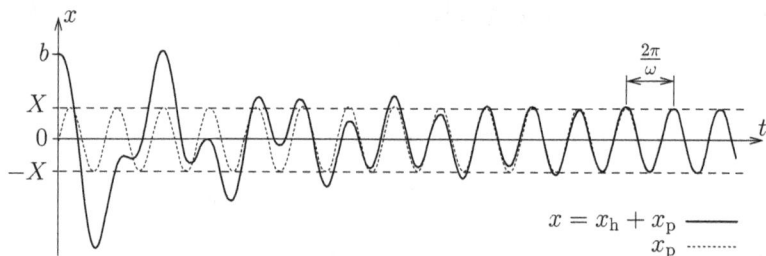

Figure 11.7: Illustration of a forced, damped oscillator, with initial conditions $x(0) = b$ and $\dot{x}(0) = 0$. After a transient, the motion is described by its particular solution (dotted line), with amplitude X.

Definition 11.1 (Magnification factor). For an oscillating system, with natural angular frequency ω_n and damping factor ζ, the *magnification factor* is

$$M(\omega) \equiv \sqrt{\frac{1}{(1 - \omega^2/\omega_n^2)^2 + (2\zeta\omega/\omega_n)^2}}, \tag{11.9}$$

where ω is the angular frequency of a periodic applied force.

When only the particular solution remains of the forced harmonic motion, its amplitude is proportional to the magnification factor $M(\omega)$:

Theorem 11.2. For an oscillating system governed by the differential equation

$$\ddot{x} + 2\zeta\omega_n\dot{x} + \omega_n^2 x = a_0 + a\sin(\omega t + \psi),$$

where $\omega_n > 0$, $\omega > 0$, $\zeta > 0$, a_0, a, and ψ are constants, the particular solution is a harmonic function with the amplitude

$$X = \frac{a}{\omega_n^2} M(\omega), \tag{11.10}$$

where $M(\omega)$ is the magnification factor.

Proof. We assume that the particular solution is a harmonic function:

$$x_p = x_0 + X\sin(\omega t + \phi),$$

where x_0, X and ϕ are constants. Substituting this assumed particular solution into the differential equation gives

$$-\omega^2 X\sin(\omega t + \phi) + 2\zeta\omega\omega_n X\cos(\omega t + \phi) + \omega_n^2 X\sin(\omega t + \phi) + \omega_n^2 x_0 = a_0 + a\sin(\omega t + \psi). \tag{11.11}$$

Since this must hold for every t, we have $\omega_n^2 x_0 = a_0$, which can be subtracted from Eq. (11.11). Subsequent division of Eq. (11.11) by $\omega_n^2 X$ yields

$$2\zeta\frac{\omega}{\omega_n}\cos(\omega t + \phi) + \left(1 - \frac{\omega^2}{\omega_n^2}\right)\sin(\omega t + \phi) = \frac{a}{\omega_n^2 X}\sin(\omega t + \psi).$$

A substitution $\theta = \omega t + \phi$ gives

$$\underbrace{2\zeta \frac{\omega}{\omega_\mathrm{n}}}_{=A} \cos\theta + \underbrace{\left(1 - \frac{\omega^2}{\omega_\mathrm{n}^2}\right)}_{=B} \sin\theta = \underbrace{\frac{a}{\omega_\mathrm{n}^2 X}}_{=C} \sin(\theta - \phi + \psi). \qquad (11.12)$$

According to Eqs. (A.6) and (A.7), this Eq. (11.12) is satisfied for some value of ϕ when $C = \sqrt{A^2 + B^2}$, giving

$$\frac{a}{\omega_\mathrm{n}^2 X} = \sqrt{(2\zeta\omega/\omega_\mathrm{n})^2 + (1 - \omega^2/\omega_\mathrm{n}^2)^2} \quad \Leftrightarrow$$

$$X = \frac{a}{\omega_\mathrm{n}^2} M(\omega). \qquad \square$$

For a harmonic force $F(t) = F_0 \sin(\omega t)$ acting on the damped system in Fig. 11.6, we identify

$$f(t) = \frac{F_0}{m} \sin(\omega t),$$

in Eq. (11.7). In Theorem 11.2, we also identify $a_0 = 0$ and $a = F_0/m$, so that the amplitude of the particular solution is given by

$$X = \frac{a}{\omega_\mathrm{n}^2} M(\omega) = \frac{\frac{F_0}{m}}{\frac{k}{m}} M(\omega) = \frac{F_0}{k} M(\omega).$$

Thus, the amplitude depends on a characteristic length F_0/k and on the magnification factor.

Resonance

The importance of the magnification factor to the amplitude of forced harmonic oscillation makes it interesting to investigate its dependence on the applied angular frequency ω. In Fig. 11.8, the graph for $M(\omega)$ is plotted for different damping ratios $\zeta = \{3, 1, 1/8, 0\}$, where $\zeta = 0$ corresponds to an undamped system.

For systems that are overdamped, $\zeta > 1$, or critically damped, $\zeta = 1$, the magnification factor decreases with the angular frequency of the applied force (Fig. 11.8). Thus, rapid oscillations are damped more efficiently than slow oscillations. In the underdamped case, $\zeta < 1$, $M(\omega)$ has a maximum at $\omega \approx \omega_\mathrm{n}$, and the amplitude of the oscillation can become very large when $\omega = \omega_\mathrm{n}$. This phenomenon is called *resonance*. Therefore, the natural angular frequency is sometimes referred to as the *resonance angular frequency*.

Vibrations

We continue our investigation of the motion of a carriage that is connected to supports *via* a spring and a damper. In this case, the carriage

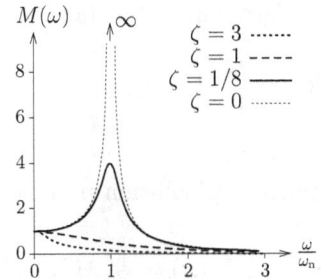

Figure 11.8: The magnification factor for different angular frequencies ω of the applied force, and for different values of ζ.

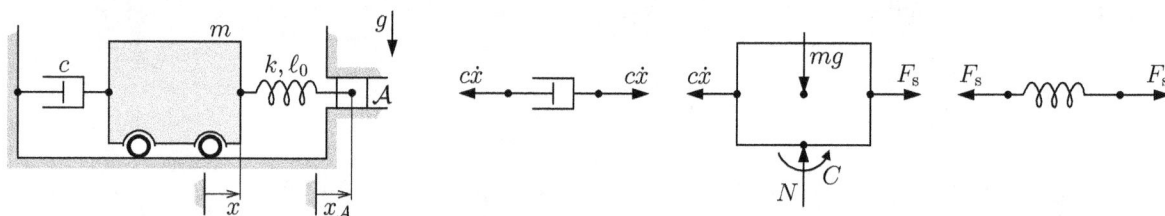

Figure 11.9: A carriage with mass m is attached to a spring and a damper. The coordinates are chosen such that the spring is undeformed when $x = x_{\mathcal{A}}$. The support $\mathcal{A}$ of the spring vibrates, so that the carriage is made to move.

is perturbed from its equilibrium position due to the prescribed motion $x_{\mathcal{A}}(t)$ of one of its supports $\mathcal{A}$ (Fig. 11.9).

Using the quantities defined in Fig. 11.9, the length of the spring is $\ell = \ell_0 + x_{\mathcal{A}} - x$, where ℓ_0 is the natural length. Therefore, the spring force is

$$F_{\mathrm{s}} = k(\ell - \ell_0) = k(x_{\mathcal{A}} - x).$$

Thus, the Law of force and acceleration applied in the x direction gives

$$\to^x: \qquad k(x_{\mathcal{A}} - x) - c\dot{x} = m\ddot{x} \quad \Leftrightarrow$$
$$\ddot{x} + \frac{c}{m}\dot{x} + \frac{k}{m}x = \frac{k}{m}x_{\mathcal{A}} \quad \Leftrightarrow$$
$$\ddot{x} + 2\zeta\omega_{\mathrm{n}}\dot{x} + \omega_{\mathrm{n}}^2 x = f(t),$$

where $f(t) = kx_{\mathcal{A}}(t)/m$. Apparently, the same type of differential equation arises when the system is set into motion by vibrations, as when a given dynamic force is applied.

For harmonic vibrations, $e.i.\ x_{\mathcal{A}}(t) = b\sin(\omega t)$ in Fig. 11.9, we have

$$f(t) = \frac{kb}{m}\sin(\omega t).$$

Then, we identify $a = kb/m$ in Theorem 11.2, so that the amplitude of the particular solution becomes

$$X = \frac{a}{\omega_{\mathrm{n}}^2}M(\omega) = \frac{\frac{kb}{m}}{\frac{k}{m}}M(\omega) = bM(\omega).$$

Thus, we can expect resonance to occur when $\omega = \omega_{\mathrm{n}}$. Once more, we observe the significance of the magnification factor to the amplitude of the oscillation.

PART III
RIGID BODY DYNAMICS

12

Planar kinematics of rigid bodies

Kinematics is the science of the geometry of motion. This chapter treats planar, rigid-body motion, that is, the motion of rigid bodies confined to a plane.

12.1 Planar motion of rigid bodies

According to Def. 1.1, a rigid body is a body where the distance between each pair of material points is unchanging. Consequently, rigid bodies cannot deform. For *planar motion*, the rigid body is further constrained.

Definition 12.1 (Planar motion). A rigid body describes *planar motion*, if there exists a *reference plane* such that every body-fixed point moves in parallel with this plane.

This means that every body-fixed point $\mathcal{P}$ of a rigid body in planar motion has a velocity $\bar{v}_{\mathcal{P}}$ and an acceleration $\bar{a}_{\mathcal{P}}$, which are both parallel with the reference plane (Fig. 12.1). If $\bar{e}_n$ denotes the unit normal of this reference plane, we have

$$\bar{v}_{\mathcal{P}} \perp \bar{e}_n, \quad \bar{a}_{\mathcal{P}} \perp \bar{e}_n.$$

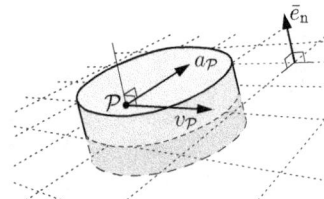

Figure 12.1: During planar motion, the velocity and acceleration of each body-fixed point $\mathcal{P}$ are parallel with the reference plane.

Translation and rotation

Two important special cases of planar motion are *translational* and *rotational* motion.

Definition 12.2 (Translational motion). A rigid body is said to describe *translational motion*, if every body-fixed point has the same velocity $\bar{v}(t)$ (Fig. 12.2ab).

Note that translational motion, or briefly *translation*, does not necessarily mean rectilinear motion. For instance, each gondola of a Ferris wheel describes translation, even though the wheel itself rotates.

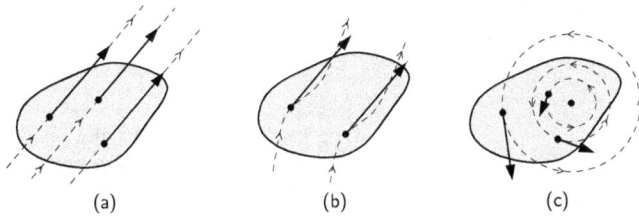

Figure 12.2: Special cases of planar motion. (a) Rectilinear translation: every body-fixed point has the same velocity, and follow straight, parallel paths. (b) Curvilinear translation: every body-fixed point has the same instantaneous velocity, even though they follow curved paths. (c) Rotation: all body-fixed points describe circular motion around an axis.

Definition 12.3 (Rotational motion). A rigid body describes *rotational motion*, if there exists an *axis of rotation*, such that every body-fixed point performs circular motion around this axis (Fig. 12.2c).

The axis of rotation can be located in or outside the body. For planar motion, the axis of rotation is perpendicular to the reference plane.

Position and orientation

Let XYZ be space-fixed coordinates with the orthogonal basis $\{\bar{e}_X, \bar{e}_Y, \bar{e}_Z\}$. Furthermore, let the XY plane be the reference plane of the planar motion of a rigid body Ω. To uniquely describe the position of the rigid body in this plane, we introduce a body-fixed, rectangular coordinate system with the coordinates x, y and z, and with its origin located in the body-fixed point $\mathcal{P}$. The basis of this body-fixed system is

$$\bar{e}_x = \cos\theta \bar{e}_X + \sin\theta \bar{e}_Y, \tag{12.1a}$$

$$\bar{e}_y = -\sin\theta \bar{e}_X + \cos\theta \bar{e}_Y, \tag{12.1b}$$

$$\bar{e}_z = \bar{e}_Z. \tag{12.1c}$$

Here, θ denotes the *polar angle* of $\bar{e}_x$, that is, the angle from the X direction to the x direction. This quantity θ represents the *orientation* of the rigid body. The position of the rigid body can be uniquely described by $\bar{r}_{\mathcal{P}}$ together with θ (Fig. 12.3). Thus, its motion can be described by the functions $\bar{r}_{\mathcal{P}}(t)$ and $\theta(t)$. Since $\theta = \theta(t)$, we have that $\bar{e}_x = \bar{e}_x(t)$ and $\bar{e}_y = \bar{e}_y(t)$ are functions of time.

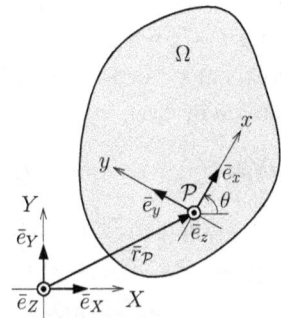

Figure 12.3: A representation of the position and orientation of a rigid body, using a space-fixed coordinate system XYZ, and a body-fixed coordinate system xyz.

Angular velocity and angular acceleration

The *angular velocity* quantifies the rate of rotation of a rigid body. Herein, we use the units of rad/s for angular velocity. However, there are other units for angular velocity. For instance, *revolutions per minute* (rpm) are commonly used for the angular velocity of parts of machinery.

Definition 12.4 (Angular velocity for planar motion). The *angular velocity* of a rigid body in planar motion is

$$\bar{\omega} \equiv \dot{\theta}\bar{e}_{\mathrm{n}}, \tag{12.2}$$

where $\bar{e}_n$ is the normal of the reference plane, and $\theta(t)$ is the polar angle of a body-fixed axis in this reference plane (Fig. 12.3).

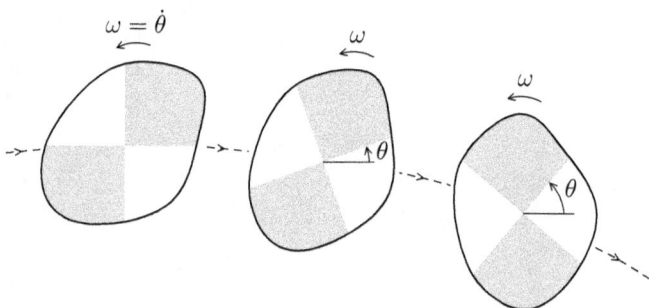

Figure 12.4: Planar motion where the angle θ to an imaginary, body-fixed line increases due to the angular velocity ω.

For planar motion, we can write $\bar{\omega} = \omega\bar{e}_n$, where $\omega = \dot{\theta}$ (Fig. 12.4). Thus, the scalar ω is sufficient to describe angular velocity in the plane. The vector direction of the angular velocity can be identified using a variant of the right-hand rule (Fig. 12.5).

Theorem 12.5. The angular velocity $\bar{\omega}$ of a rigid body in planar motion is independent of the choice of body-fixed coordinate system.

Proof. We choose the XY plane as the reference plane, and consider two different body-fixed coordinate systems: xyz with origin $\mathcal{P}$, and $x^*y^*z^*$ with origin $\mathcal{Q}$, where $\bar{e}_z = \bar{e}_{z^*} = \bar{e}_Z$. The angle ϕ from $\bar{e}_x$ to $\bar{e}_{x^*}$ is constant, since both these unit vectors are body-fixed (Fig. 12.6). If the polar angle of $\bar{e}_x$ is $\theta(t)$, then the polar angle of $\bar{e}_{x^*}$ is

$$\theta^*(t) = \theta(t) + \phi \quad \Rightarrow \quad \dot{\theta}^* = \dot{\theta}.$$

Thus, according to Def. 12.4, both choices of coordinate system give the same angular velocity $\bar{\omega} = \dot{\theta}\bar{e}_Z$. $\qquad\square$

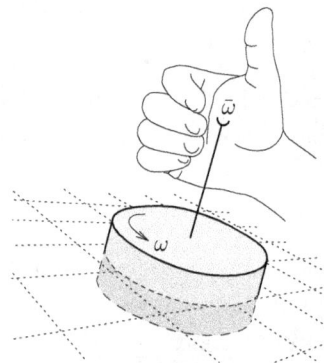

Figure 12.5: The vector direction of the angular velocity is identified by aligning the fingers of the right hand, except the thumb, with the rotation direction. The thumb then points in the direction of the angular velocity vector.

Since the angular velocity $\bar{\omega}$ is independent of the position of the body-fixed coordinate system, the angular velocity is a free vector, unrelated to any distinct point within the body.

Definition 12.6 (Angular acceleration for planar motion). The *angular acceleration* of a rigid body in planar motion, with angular velocity $\bar{\omega}$, is

$$\bar{\alpha} \equiv \dot{\bar{\omega}} = \ddot{\theta}\bar{e}_n, \tag{12.3}$$

where $\bar{e}_n$ is the normal of the reference plane, and $\theta(t)$ is the polar angle of a body-fixed axis in this reference plane.

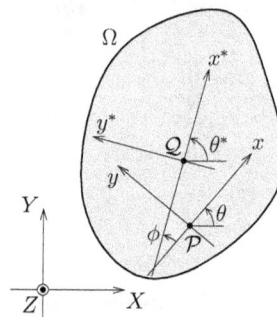

Figure 12.6: Geometry of Theorem 12.5. Two body-fixed coordinate systems, xyz and $x^*y^*z^*$, with different orientations and origins, are introduced.

The angular acceleration is defined through the angular velocity, and the angular velocity is a free vector. Therefore, the angular acceleration $\bar{\alpha}$ is also a free vector.

Relative velocity and acceleration

Lemma 12.7 (Time derivative of a body-fixed vector). For a rigid body in planar motion with angular velocity $\bar{\omega}$, the time derivative of a body-fixed vector $\bar{u}$ is

$$\dot{\bar{u}} = \bar{\omega} \times \bar{u}. \tag{12.4}$$

Proof. We choose a space-fixed coordinate system XYZ, such that the XY plane is the reference plane. Furthermore, we introduce a body-fixed coordinate system xyz, where $\bar{e}_z = \bar{e}_Z$. Thus, the vector $\bar{u}$ can be written

$$\bar{u} = u_x \bar{e}_x + u_y \bar{e}_y, \tag{12.5}$$

where u_x and u_y are constant components (Fig. 12.7). If the polar angle of $\bar{e}_x$ is $\theta(t)$, then, by inserting Eq. (12.1a) and Eq. (12.1b) into Eq. (12.5), we obtain

$$\bar{u} = u_x(\cos\theta \bar{e}_X + \sin\theta \bar{e}_Y) + u_y(-\sin\theta \bar{e}_X + \cos\theta \bar{e}_Y).$$

Differentiation of this equation w.r.t. time gives an expression for the left-hand side of Eq. (12.4):

$$\dot{\bar{u}} = u_x\dot{\theta}(-\sin\theta \bar{e}_X + \cos\theta \bar{e}_Y) + u_y\dot{\theta}(-\cos\theta \bar{e}_X - \sin\theta \bar{e}_Y)$$
$$= \dot{\theta}(u_x \bar{e}_y - u_y \bar{e}_x).$$

Moreover, if we use Def. 12.4 for the angular velocity, and Eq. (12.5), to calculate the right-hand side of Eq. (12.4), we obtain

$$\bar{\omega} \times \bar{u} = \dot{\theta}\bar{e}_z \times (u_x\bar{e}_x + u_y\bar{e}_y)$$
$$= \dot{\theta}\left[u_x(\bar{e}_z \times \bar{e}_x) + u_y(\bar{e}_z \times \bar{e}_y)\right]$$
$$= \dot{\theta}(u_x\bar{e}_y - u_y\bar{e}_x).$$

Therefore, the left-hand and right-hand sides of Eq. (12.4) are equal. $\square$

Theorem 12.8 (Relative-velocity equation). For a rigid body in planar motion with angular velocity $\bar{\omega}$, it holds that

$$\bar{v}_{\mathcal{Q}} = \bar{v}_{\mathcal{P}} + \bar{\omega} \times \overline{\mathcal{PQ}}, \tag{12.6}$$

where $\mathcal{P}$ and $\mathcal{Q}$ are body-fixed points.

Proof. The Parallelogram law gives (Fig. 12.8)

$$\bar{r}_{\mathcal{Q}} = \bar{r}_{\mathcal{P}} + \overline{\mathcal{PQ}}.$$

Time differentiation, while observing that $\overline{\mathcal{PQ}}$ is a body-fixed vector, gives

$$\dot{\bar{r}}_{\mathcal{Q}} = \dot{\bar{r}}_{\mathcal{P}} + \dot{\overline{\mathcal{PQ}}} \quad \Leftrightarrow \quad \{\text{Theorem 12.7}\} \quad \Leftrightarrow$$
$$\bar{v}_{\mathcal{Q}} = \bar{v}_{\mathcal{P}} + \bar{\omega} \times \overline{\mathcal{PQ}}.$$

$\square$

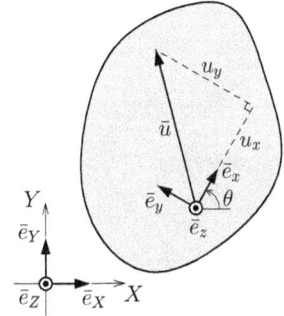

Figure 12.7: Geometry for the proof of Lemma 12.7.

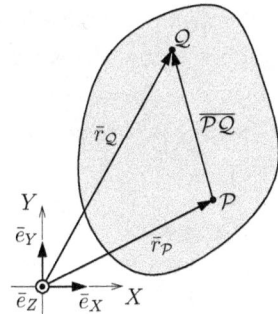

Figure 12.8: Geometry for the proof of Theorem 12.8.

Every rigid body in a mechanical system has its own angular velocity. If we know the velocity of a body-fixed point and the angular velocity of the body, we can calculate the velocity of any other body-fixed point within this body by using Eq. (12.6). Differentiation of Eq. (12.6) w.r.t. time gives a relation between the accelerations of different body-fixed points.

Theorem 12.9 (Relative-acceleration equation). For a rigid body in planar motion with angular velocity $\bar{\omega}$ and angular acceleration $\bar{\alpha}$, it holds that

$$\bar{a}_Q = \bar{a}_P + \bar{\alpha} \times \overline{PQ} + \bar{\omega} \times (\bar{\omega} \times \overline{PQ}), \qquad (12.7)$$

where P and Q are body-fixed points.

Proof. Differentiation of Eq. (12.6) w.r.t. time gives

$$\dot{\bar{v}}_Q = \dot{\bar{v}}_P + \frac{\mathrm{d}}{\mathrm{d}t}\left(\bar{\omega} \times \overline{PQ}\right) \quad \Leftrightarrow \quad \{\text{Product rule (A.25c)}\} \quad \Leftrightarrow$$

$$\bar{a}_Q = \bar{a}_P + \dot{\bar{\omega}} \times \overline{PQ} + \bar{\omega} \times \dot{\overline{PQ}}.$$

According to Def. 12.6, $\bar{\alpha} = \dot{\bar{\omega}}$, and according to Lemma 12.7, $\dot{\overline{PQ}} = \bar{\omega} \times \overline{PQ}$, which give

$$\bar{a}_Q = \bar{a}_P + \bar{\alpha} \times \overline{PQ} + \bar{\omega} \times (\bar{\omega} \times \overline{PQ}). \qquad \square$$

12.2   Instantaneous center

According to Eq. (12.6), the velocity varies between different points within a rigid body. Let us ask a question: Could it be, that for *every* rigid-body motion, it is possible to find a body-fixed point[21] with zero velocity?

Theorem 12.10 (Instantaneous center). For a rigid body in planar motion, with angular velocity $\bar{\omega} \neq \bar{0}$, there exists in every moment a unique body-fixed point C in the reference plane, called the *instantaneous center*, such that $\bar{v}_C = \bar{0}$.

Proof. Let xyz be a space-fixed coordinate system, with the xy plane as the reference plane, so that $\bar{\omega} = \omega \bar{e}_z$. Choose a body-fixed point P with position vector $\bar{r}_P = x_P \bar{e}_x + y_P \bar{e}_y$ and velocity $\bar{v}_P$. Every body-fixed point C, with position vector $\bar{r}_C = x_C \bar{e}_x + y_C \bar{e}_y$ and velocity $\bar{v}_C = \bar{0}$, must satisfy Eq. (12.6):

$$\bar{0} = \bar{v}_P + \bar{\omega} \times \overline{PC} \quad \Leftrightarrow$$

$$\begin{cases} 0 = v_{Px} - \omega(y_C - y_P) \\ 0 = v_{Py} + \omega(x_C - x_P) \end{cases} \quad \Leftrightarrow$$

[21] *Body-fixed point* – a point that is fixed relative to a body-fixed coordinate system. However, the point is not required to be located within the body.

$$\underbrace{\begin{bmatrix} 0 & \omega \\ -\omega & 0 \end{bmatrix}}_{=\bar{\bar{A}}} \begin{bmatrix} x_{\mathcal{C}} \\ y_{\mathcal{C}} \end{bmatrix} = \begin{bmatrix} v_{\mathcal{P}x} + \omega y_{\mathcal{P}} \\ v_{\mathcal{P}y} - \omega x_{\mathcal{P}} \end{bmatrix}.$$

Since $\det \bar{\bar{A}} = \omega^2 \neq 0$,[22] there is a unique solution $(x_{\mathcal{C}}, y_{\mathcal{C}})$. Therefore, the body-fixed point $\mathcal{C}$, such that $\bar{v}_{\mathcal{C}} = \bar{0}$, is unique. □

[22] Matrices are denoted by a double-bar over the variable name.

Consequently, in every instant there exists exactly *one* body-fixed point $\mathcal{C}$ inside or outside the body, that has zero velocity. In this instant, it appears as though all body-fixed points describe circular motion around $\mathcal{C}$.

If $\mathcal{C}$ denotes the instantaneous center of a rigid body, it follows from Eq. (12.6) that the velocity of a body-fixed point $\mathcal{P}$ is

$$\bar{v}_{\mathcal{P}} = \bar{\omega} \times \overline{\mathcal{C}\mathcal{P}}.$$

The properties of the cross product ensure that the velocity $\bar{v}_{\mathcal{P}}$ is perpendicular to the line $\mathcal{C}\mathcal{P}$, and that

$$v_{\mathcal{P}} = \pm \omega r, \qquad r = |\overline{\mathcal{C}\mathcal{P}}|. \tag{12.8}$$

Thus, the speed of body-fixed points increases linearly with the distance r from the instantaneous center.

The instantaneous center of a rigid body can be constructed graphically, if the velocity directions are known for at least two body-fixed points, $\mathcal{P}$ and $\mathcal{Q}$. Draw a line $\mathcal{L}_1$ perpendicular to $\bar{v}_{\mathcal{P}}$ through $\mathcal{P}$, and a line $\mathcal{L}_2$ perpendicular to $\bar{v}_{\mathcal{Q}}$ through $\mathcal{Q}$. The instantaneous center $\mathcal{C}$ is the intersection of $\mathcal{L}_1$ and $\mathcal{L}_2$ (Fig. 12.9). Using Eq. (12.8), we obtain the angular velocity $\omega = v_{\mathcal{P}}/|\overline{\mathcal{C}\mathcal{P}}|$.

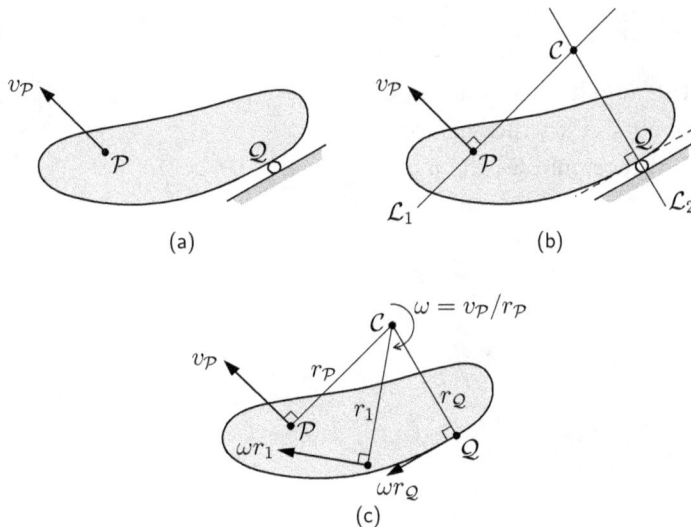

Figure 12.9: (a) A rigid body has a known velocity in $\mathcal{P}$, and a known velocity direction in $\mathcal{Q}$. (b) Draw a line $\mathcal{L}_1$ perpendicular to the velocity direction of $\mathcal{P}$, and a line $\mathcal{L}_2$ perpendicular to the velocity direction of $\mathcal{Q}$. The intersection of these lines is the instantaneous center $\mathcal{C}$. (c) The angular velocity is given by Eq. (12.8), and the velocity of each point on the body is perpendicular to the ray from $\mathcal{C}$ to this point.

Another important example is fixed-axis rotation, where the point of the axis of rotation *is* the instantaneous center $\mathcal{C}$, and every body-fixed point describes circular motion around $\mathcal{C}$ (Fig. 12.10).

12.3   Rolling without slipping

If a rigid body with a rounded shape rolls against a surface without slipping, then the velocity difference between the body and the surface is zero at the contact point. Therefore, this contact point is the instantaneous center $\mathcal{C}$ of the rolling rigid body. A rolling wheel presents an important special case.

Consider a wheel with radius R, that rolls without slipping on a plane surface. The hub $\mathcal{P}$ of the wheel will move rectilinearly and parallel to the surface. The velocity of the contact point $\mathcal{C}$ between the wheel and the surface is $\bar{v}_{\mathcal{C}} = \bar{0}$. Thus, this contact point is the instantaneous center of the wheel. According to Eq. (12.8), the hub has the velocity

$$v_{\mathcal{P}} = \omega|\overline{\mathcal{CP}}| = \omega R, \tag{12.9}$$

directed along the surface, where ω is the angular velocity of the wheel (Fig. 12.11). Moreover, the velocity of every body-fixed point $\mathcal{Q}$ on the wheel is perpendicular to a ray $\mathcal{CQ}$ (Fig. 12.12), and

$$v_{\mathcal{Q}} = \omega|\overline{\mathcal{CQ}}|. \tag{12.10}$$

Since the hub $\mathcal{P}$ of the wheel describes rectilinear motion (Fig. 12.11), its acceleration is given by Eq. (7.2),

$$\begin{aligned} a_{\mathcal{P}} = \dot{v}_{\mathcal{P}} &= \big\{\text{Eq. (12.9)}\big\} \\ &= \dot{\omega}R \\ &= \alpha R, \end{aligned}$$

directed along the surface, where $\alpha = \dot{\omega}$ is the angular acceleration of the wheel. The acceleration in the contact point $\mathcal{C}$ with the surface is nonzero. Let the x axis be directed along the surface, and let the y axis an upward surface normal, so that (Fig. 12.13)

$$\bar{a}_{\mathcal{P}} = \alpha R \bar{e}_x, \quad \bar{\omega} = -\omega \bar{e}_z, \quad \bar{\alpha} = -\alpha \bar{e}_z.$$

The Relative-acceleration equation (12.7) gives

$$\begin{aligned} \bar{a}_{\mathcal{C}} &= \bar{a}_{\mathcal{P}} + \bar{\alpha} \times \overline{\mathcal{PC}} + \bar{\omega} \times (\bar{\omega} \times \overline{\mathcal{PC}}) \\ &= \underbrace{\alpha R \bar{e}_x - \alpha \bar{e}_z \times R(-\bar{e}_y)}_{=\bar{0}} - \omega \bar{e}_z \times [-\omega \bar{e}_z \times R(-\bar{e}_y)] \\ &= \omega^2 R \bar{e}_y. \end{aligned}$$

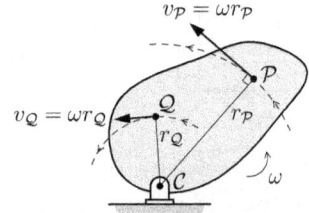

Figure 12.10: During fixed-axis rotation, the instantaneous center $\mathcal{C}$ is located on the axis of rotation.

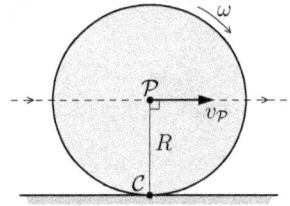

Figure 12.11: A wheel rolling towards the right without slipping on the plane surface. The hub $\mathcal{P}$ moves rectilinearly. The contact point $\mathcal{C}$ is the instantaneous center.

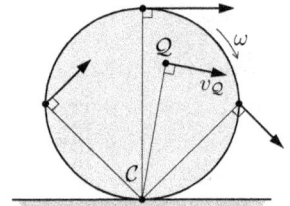

Figure 12.12: A wheel rolling towards the right without slipping on the plane surface. The velocity of each body-fixed point $\mathcal{Q}$ is perpendicular to a ray $\mathcal{CQ}$.

Thus, at the instantaneous center, the acceleration of the body-fixed point is directed towards the hub of the wheel. This can be understood from the fact that the body-fixed point is at a turning point in its path (Fig. 12.13).

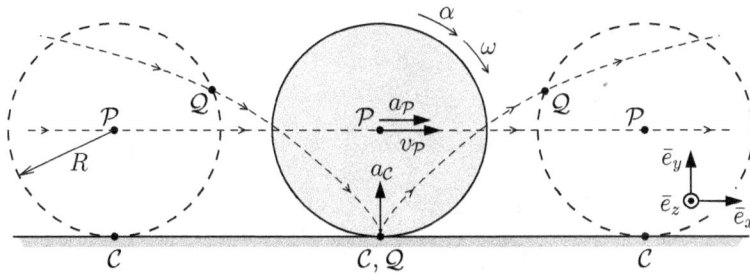

Figure 12.13: A wheel rolls to the right without slipping on the plane surface. A point $\mathcal{Q}$ on the periphery of the wheel moves along a path with its turning point at the instantaneous center $\mathcal{C}$.

13
Planar kinetics of rigid bodies

Planar kinetics is the study of motion of rigid bodies confined to a plane. Some of the definitions and theorems in this chapter also hold for general motion in three dimensions. If a theorem is restricted to planar motion, then this is clearly stated in its formulation.

13.1 Euler's laws of motion

While Newton's laws of motion are valid for particles, *Euler's laws of motion* are valid for deformable bodies translating and rotating in space. They describe how a force-couple system affects the momentum and angular momentum of a body. We begin by defining the momentum and angular momentum of general, deformable bodies.

Definition 13.1 (Momentum). The *momentum* of a body Ω is

$$\bar{G} \equiv \int_\Omega \bar{v}\,\mathrm{d}m, \tag{13.1}$$

where $\bar{v}$ denotes the velocity of the mass element $\mathrm{d}m$ in an inertial system (Fig. 13.1).

Definition 13.2 (Angular momentum). The *angular momentum* of a body Ω, w.r.t. an arbitrary point $\mathcal{A}$, is

$$\bar{H}_\mathcal{A} \equiv \int_\Omega (\bar{r} \times \bar{v})\,\mathrm{d}m, \tag{13.2}$$

where $\bar{r}$ is a vector from $\mathcal{A}$ to the mass element $\mathrm{d}m$, and $\bar{v}$ is the velocity of this mass element in an inertial system (Fig. 13.1).

If a body is affected by a force-couple system (Def. 2.7), with force sum $\Sigma \bar{F}$ (Def. 2.8) and moment sum $\Sigma \bar{M}_\mathcal{D}$ w.r.t. a space-fixed point $\mathcal{D}$[23] (Def. 2.9), then this force-couple system yields a change in the momentum and angular momentum of the body, according to Euler's laws of motion:

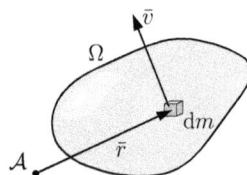

Figure 13.1: Geometry used in Defs. 13.1 and 13.2, where $\bar{r}$ is a vector from an arbitrary point $\mathcal{A}$ to the mass element $\mathrm{d}m$, and where $\bar{v}$ is the velocity of this mass element.

[23] *Space-fixed point* – a fixed point in a given inertial system.

Postulate 13.3 (Euler's laws of motion). For a body affected by a force-couple system with force sum $\Sigma \bar{F}$ and moment sum $\Sigma \bar{M}_{\mathcal{D}}$ w.r.t. a space-fixed point $\mathcal{D}$, it holds that

$$\Sigma \bar{F} = \dot{\bar{G}}, \tag{13.3a}$$

$$\Sigma \bar{M}_{\mathcal{D}} = \dot{\bar{H}}_{\mathcal{D}}, \tag{13.3b}$$

where $\bar{G}$ is the momentum, and $\bar{H}_{\mathcal{D}}$ is the angular momentum of the body w.r.t. $\mathcal{D}$.

Euler's first law, Eq. (13.3a), is called the *Force equation*, and describes how the translation of the body is affected by the force-couple system. Euler's second law, Eq. (13.3b), is called the *Moment equation* and describes how the rotation of the body is affected by the force-couple system. Observe that Euler's laws are only valid in inertial systems.

Below, we investigate how the expressions for momentum and angular momentum can be simplified for *rigid* bodies, and how Euler's laws of motion can be used and interpreted.

13.2   Euler's first law: the Force equation

The expression for momentum can be simplified for rigid bodies. This, in turn, simplifies the Force equation. In order to derive these relations, we need to formulate a *lemma*.

Lemma 13.4. For a rigid body Ω, it holds that

$$\int_{\Omega} \bar{s}\,\mathrm{d}m = \bar{0}, \tag{13.4}$$

where $\bar{s}$ is a geometric vector from the center of mass $\mathcal{G}$ of Ω to the mass element $\mathrm{d}m$.

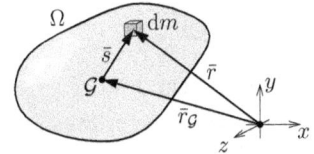

Figure 13.2: A rigid body with a vector $\bar{s}$ from the center of mass $\mathcal{G}$ to a mass element $\mathrm{d}m$. Here, $\bar{r}$ is the position vector of the mass element relative to a given coordinate system.

Proof. According to the Parallelogram law, the position vector of the mass element is $\bar{r} = \bar{r}_{\mathcal{G}} + \bar{s}$ (Fig. 13.2). If $m > 0$ denotes the mass of the rigid body, then Def. 4.1 gives

$$
\begin{aligned}
\bar{r}_{\mathcal{G}} &= \frac{1}{m} \int_{\Omega} \bar{r}\,\mathrm{d}m \\
&= \frac{1}{m} \int_{\Omega} (\bar{r}_{\mathcal{G}} + \bar{s})\,\mathrm{d}m = \{\bar{r}_{\mathcal{G}} \text{ const.}\} \\
&= \frac{1}{m} \bar{r}_{\mathcal{G}} \underbrace{\int_{\Omega} \mathrm{d}m}_{=m} + \frac{1}{m} \int_{\Omega} \bar{s}\,\mathrm{d}m \\
&= \bar{r}_{\mathcal{G}} + \frac{1}{m} \int_{\Omega} \bar{s}\,\mathrm{d}m.
\end{aligned}
$$

Subtracting $\bar{r}_{\mathcal{G}}$ from both sides yields Eq. (13.4).   □

Theorem 13.5 (Momentum of a rigid body). The momentum of a rigid body is

$$\bar{G} = m\bar{v}_\mathcal{G}, \tag{13.5}$$

where m is the mass, and $\bar{v}_\mathcal{G}$ is the velocity of the center of mass $\mathcal{G}$ of the body.

Proof. Let $\bar{r}_\mathcal{G}$ denote the position of $\mathcal{G}$, so that the position vector $\bar{r}$ of a mass element can be written

$$\bar{r} = \bar{r}_\mathcal{G} + \bar{s} \quad \Rightarrow \quad \dot{\bar{r}} = \dot{\bar{r}}_\mathcal{G} + \dot{\bar{s}} \quad \Leftrightarrow \quad \bar{v} = \bar{v}_\mathcal{G} + \dot{\bar{s}},$$

where $\bar{s}$ is a vector originating from $\mathcal{G}$ (Fig. 13.2). Thus, Def. 13.1 applied to the body Ω gives

$$\begin{aligned}
\bar{G} &= \int_\Omega \bar{v}\,\mathrm{d}m \\
&= \int_\Omega \left(\bar{v}_\mathcal{G} + \dot{\bar{s}} \right) \mathrm{d}m = \{\bar{v}_\mathcal{G} \text{ const.}\} \\
&= \bar{v}_\mathcal{G} \underbrace{\int_\Omega \mathrm{d}m}_{=m} + \int_\Omega \dot{\bar{s}}\,\mathrm{d}m \\
&= m\bar{v}_\mathcal{G} + \frac{\mathrm{d}}{\mathrm{d}t}\left(\int_\Omega \bar{s}\,\mathrm{d}m \right) = \{\text{Lemma 13.4}\} \\
&= m\bar{v}_\mathcal{G}. \qquad\qquad\qquad\qquad\qquad\qquad\quad \square
\end{aligned}$$

Time differentiation of Eq. (13.5) gives $\dot{\bar{G}} = m\bar{a}_\mathcal{G}$. Substitution of this equation into the Force equation, Eq. (13.3a), gives

$$\Sigma\bar{F} = m\bar{a}_\mathcal{G}. \tag{13.6}$$

Consequently, the law of motion for the center of mass of a rigid body takes a similar expression as the Law of force and acceleration, Eq. (8.1), for a particle.

13.3 Euler's second law: the Moment equation

The Moment equation (13.3b) is formulated for the angular momentum w.r.t. a space-fixed point $\mathcal{D}$. We will show that the Moment equation can be similarly formulated w.r.t. the center of mass $\mathcal{G}$ of a rigid body. For this purpose, the Transfer theorem of angular momentum is required.

Theorem 13.6 (Transfer theorem of angular momentum). For a body with center of mass $\mathcal{G}$, and for an arbitrary point $\mathcal{A}$, it holds that

$$\bar{H}_\mathcal{A} = \bar{H}_\mathcal{G} + \overline{\mathcal{AG}} \times \bar{G}, \tag{13.7}$$

where $\bar{H}_\mathcal{A}$ and $\bar{H}_\mathcal{G}$ are the angular momenta of the body w.r.t. $\mathcal{A}$ and $\mathcal{G}$, respectively, and where $\bar{G}$ is the momentum of the body.

Proof. If we place the origin in $\mathcal{A}$, and let $\bar{r}_{\mathcal{G}} = \overline{\mathcal{AG}}$, then the position vector $\bar{r}$ of a mass element $\mathrm{d}m$ in the rigid body Ω can be written

$$\bar{r} = \bar{r}_{\mathcal{G}} + \bar{s},$$

where $\bar{s}$ is a vector originating from $\mathcal{G}$ (Fig. 13.2). Definition 13.2 concerning angular momentum gives

$$
\begin{aligned}
\bar{H}_{\mathcal{A}} &= \int_{\Omega} (\bar{r} \times \bar{v}) \mathrm{d}m \\
&= \int_{\Omega} \left[(\bar{r}_{\mathcal{G}} + \bar{s}) \times \bar{v} \right] \mathrm{d}m \\
&= \int_{\Omega} (\bar{r}_{\mathcal{G}} \times \bar{v}) \, \mathrm{d}m + \int_{\Omega} (\bar{s} \times \bar{v}) \, \mathrm{d}m = \left\{ \text{Def. 13.2, w.r.t. } \mathcal{G} \right\} \\
&= \int_{\Omega} (\bar{r}_{\mathcal{G}} \times \bar{v}) \, \mathrm{d}m + \bar{H}_{\mathcal{G}} = \left\{ \bar{r}_{\mathcal{G}} \text{ const.} \right\} \\
&= \bar{r}_{\mathcal{G}} \times \int_{\Omega} \bar{v} \mathrm{d}m + \bar{H}_{\mathcal{G}} = \left\{ \text{Def. 13.1} \right\} \\
&= \overline{\mathcal{AG}} \times \bar{G} + \bar{H}_{\mathcal{G}}. \qquad\qquad \square
\end{aligned}
$$

Thus, for rigid bodies, Theorems 13.5 and 13.6 give that

$$\bar{H}_{\mathcal{A}} = \bar{H}_{\mathcal{G}} + \overline{\mathcal{AG}} \times m\bar{v}_{\mathcal{G}}. \tag{13.8}$$

We will use this result to rewrite the Moment equation, so that it can be applied w.r.t. the center of mass $\mathcal{G}$.

Theorem 13.7 (Moment equation w.r.t. the center of mass). For a rigid body affected by a force-couple system with moment sum $\Sigma\bar{M}_{\mathcal{G}}$, w.r.t. the center of mass $\mathcal{G}$, it holds that

$$\Sigma\bar{M}_{\mathcal{G}} = \dot{\bar{H}}_{\mathcal{G}}, \tag{13.9}$$

where $\bar{H}_{\mathcal{G}}$ is the angular momentum of the body w.r.t. $\mathcal{G}$.

Proof. We place the origin in a space-fixed point $\mathcal{D}$, and let $\bar{r}_{\mathcal{G}} = \overline{\mathcal{DG}}$. The Transfer theorem of the moment sum, Eq. 2.10, gives

$$
\begin{aligned}
\Sigma\bar{M}_{\mathcal{D}} &= \Sigma\bar{M}_{\mathcal{G}} + \overline{\mathcal{DG}} \times \Sigma\bar{F} = \left\{ \text{Force equation (13.6)} \right\} \\
&= \Sigma\bar{M}_{\mathcal{G}} + \bar{r}_{\mathcal{G}} \times m\bar{a}_{\mathcal{G}}. \tag{13.10}
\end{aligned}
$$

Differentiation of the Transfer theorem for angular momentum 13.6 w.r.t. time gives

$$
\begin{aligned}
\dot{\bar{H}}_{\mathcal{D}} &= \frac{\mathrm{d}}{\mathrm{d}t} \left(\bar{H}_{\mathcal{G}} + \overline{\mathcal{DG}} \times \bar{G} \right) = \left\{ \text{Theorem 13.5} \right\} \\
&= \dot{\bar{H}}_{\mathcal{G}} + \frac{\mathrm{d}}{\mathrm{d}t} (\bar{r}_{\mathcal{G}} \times m\bar{v}_{\mathcal{G}}) = \left\{ \text{Product rule (A.25c)} \right\} \\
&= \dot{\bar{H}}_{\mathcal{G}} + \bar{v}_{\mathcal{G}} \times m\bar{v}_{\mathcal{G}} + \bar{r}_{\mathcal{G}} \times m\bar{a}_{\mathcal{G}} = \left\{ \bar{v}_{\mathcal{G}} \times \bar{v}_{\mathcal{G}} = \bar{0} \right\} \\
&= \dot{\bar{H}}_{\mathcal{G}} + \bar{r}_{\mathcal{G}} \times m\bar{a}_{\mathcal{G}}. \tag{13.11}
\end{aligned}
$$

Substitution of Eqs. (13.10) and (13.11) into Eq. (13.3b) gives

$$\Sigma \bar{M}_{\mathcal{G}} = \dot{\bar{H}}_{\mathcal{G}}. \qquad \qquad \square$$

13.4   Flat rigid bodies in planar motion

Previously in this chapter, we established a theoretical framework for rigid bodies of arbitrary shape in general, three-dimensional motion. Next, we restrict our study to flat rigid bodies in planar motion.

Definition 13.8 (Flat rigid body). A *flat* rigid body has a symmetric geometry and a symmetric density w.r.t. a symmetry plane, whose normal is called the *thickness direction* (Fig. 13.3a).

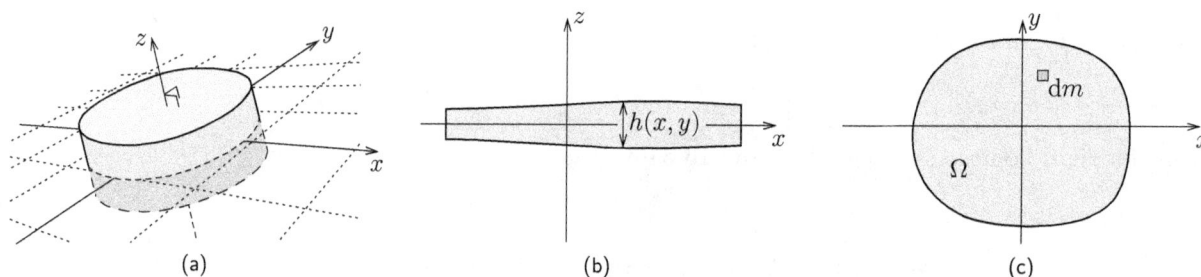

(a)   (b)   (c)

Figure 13.3: (a) A flat rigid body with the xy plane as its symmetry plane. (b) Cross–section of a flat body. The height $h(x, y)$ is allowed to vary across the plane. (c) The projected surface Ω of a flat body, having a surface element dA and mass element dm.

The thickness h of the flat rigid body is its extension in the thickness direction (Fig. 13.3b), and the *surface density* of the flat rigid body is defined as

$$\varrho_A \equiv \int_{-h/2}^{h/2} \varrho \, dz, \qquad (13.12)$$

where ϱ is the density, and z is a position coordinate in the thickness direction with its origin in the symmetry plane. For a uniform density, it holds that $\varrho_A = h\varrho$. For a flat rigid body Ω, the mass element is $dm = \varrho_A dA$, so that the mass of the body is given by

$$m = \int_\Omega dm = \int_\Omega \varrho_A \, dA, \qquad (13.13)$$

where dA is a surface element in the symmetry plane (Fig. 13.3c).

Moment of inertia

The mass of a rigid body quantifies its resistance to changing the velocity of its center of mass. This phenomenon is called the *inertia* of the body. Similarly, a flat rigid body offers resistance to changing its angular velocity. This resistance is called the *moment of inertia*.[24]

---

[24] Also called *mass moment of inertia*.

Definition 13.9 (Moment of inertia). The moment of inertia of a flat rigid body Ω, w.r.t. an arbitrary point $\mathcal{A}$ in the symmetry plane, is

$$I_{\mathcal{A}} \equiv \int_{\Omega} (\bar{r} \cdot \bar{r}) \mathrm{d}m = \int_{\Omega} r^2 \mathrm{d}m, \tag{13.14}$$

where $\bar{r}$ is the vector from $\mathcal{A}$ to the mass element $\mathrm{d}m$.

Particularly, according to Def. 13.9, the moment of inertia w.r.t. the center of mass $\mathcal{G}$ is given by

$$I_{\mathcal{G}} = \int_{\Omega} (\bar{s} \cdot \bar{s}) \mathrm{d}m = \int_{\Omega} s^2 \mathrm{d}m, \tag{13.15}$$

where $\bar{s}$ is a vector from $\mathcal{G}$ to the mass element (Fig. 13.4). If $I_{\mathcal{G}}$ is given, it is possible to determine the moment of inertia w.r.t. an arbitrary point $\mathcal{A}$ using *Steiner's theorem*.

Theorem 13.10 (Steiner's theorem). The moment of inertia of a rigid body, w.r.t. an arbitrary point $\mathcal{A}$, is

$$I_{\mathcal{A}} = I_{\mathcal{G}} + m|\overline{\mathcal{AG}}|^2, \tag{13.16}$$

where $I_{\mathcal{G}}$ is the moment of inertia w.r.t. the center of mass $\mathcal{G}$, and m is the mass of the body.

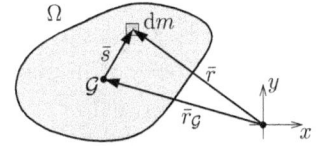

Figure 13.4: A flat rigid body with a vector $\bar{s}$ from the center of mass $\mathcal{G}$ to the mass element $\mathrm{d}m$. The vector $\bar{r}$ is the position vector of $\mathrm{d}m$.

Proof. Choose a coordinate system with its origin in $\mathcal{A}$. Let $\bar{r}_{\mathcal{G}} = \overline{\mathcal{AG}}$, so that the position vector $\bar{r}$ of the mass element can be written $\bar{r} = \bar{r}_{\mathcal{G}} + \bar{s}$, where $\bar{s}$ is a vector originating from $\mathcal{G}$ (Fig. 13.4). For the rigid body Ω, Def. 13.9 gives

$$\begin{aligned}
I_{\mathcal{A}} &= \int_{\Omega} (\bar{r} \cdot \bar{r}) \mathrm{d}m \\
&= \int_{\Omega} [(\bar{r}_{\mathcal{G}} + \bar{s}) \cdot (\bar{r}_{\mathcal{G}} + \bar{s})] \, \mathrm{d}m = \{\text{Eq. (A.16b)}\} \\
&= \int_{\Omega} (\bar{r}_{\mathcal{G}} \cdot \bar{r}_{\mathcal{G}} + 2\bar{r}_{\mathcal{G}} \cdot \bar{s} + \bar{s} \cdot \bar{s}) \, \mathrm{d}m = \{\bar{r}_{\mathcal{G}} \text{ const.}\} \\
&= |\bar{r}_{\mathcal{G}}|^2 \underbrace{\int_{\Omega} \mathrm{d}m}_{=m} + 2\bar{r}_{\mathcal{G}} \cdot \int_{\Omega} \bar{s} \mathrm{d}m + \underbrace{\int_{\Omega} (\bar{s} \cdot \bar{s}) \mathrm{d}m}_{=I_{\mathcal{G}}} = \{\text{Lemma 13.4}\} \\
&= m|\overline{\mathcal{AG}}|^2 + I_{\mathcal{G}}. \qquad \qquad \qquad \square
\end{aligned}$$

Angular momentum

For flat rigid bodies in planar motion, Eq. (13.2) for the angular momentum $\bar{H}_{\mathcal{A}}$ w.r.t. an arbitrary point $\mathcal{A}$ can be simplified. To avoid elaborate derivations, we assume that the thickness h of the body is small as compared to the characteristic length of the projected surface.[25] Let $\bar{e}_{\mathrm{n}}$ be

[25] The theorems for flat rigid bodies in this chapter can be derived without the limitation that the thickness is small.

the normal direction of the reference plane. From Eq. (13.2), using that $\bar{r} \perp \bar{e}_n$ and $\bar{v} \perp \bar{e}_n$ (Fig. 13.5), we have

$$\bar{H}_{\mathcal{A}} = \int_{\Omega} (\bar{r} \times \bar{v}) \mathrm{d}m = \{\bar{r} \times \bar{v} \,\|\, \bar{e}_n\} = H_{\mathcal{A}} \bar{e}_n, \tag{13.17}$$

for some scalar $H_{\mathcal{A}}$. Consequently, for planar problems, the angular momentum can be represented by a scalar $H_{\mathcal{A}}$, while it is implicit that its vector direction is $\bar{e}_n$. This is analogous to the convention that an angular velocity $\bar{\omega} = \omega \bar{e}_n$ is represented by a scalar ω for planar motion. Expressions for the angular momentum in planar problems are derived below.

Lemma 13.11. The angular momentum w.r.t. the center of mass $\mathcal{G}$ of a flat rigid body Ω in planar motion is

$$\bar{H}_{\mathcal{G}} = \int_{\Omega} \bar{s} \times (\bar{\omega} \times \bar{s}) \mathrm{d}m, \tag{13.18}$$

where $\bar{s}$ is a vector from $\mathcal{G}$ to the mass element, and $\bar{\omega}$ is the angular velocity of the body.

Proof. Let $\bar{r}_{\mathcal{G}}$ be the position vector of $\mathcal{G}$, so that the position vector of the mass element can be written as (Fig. 13.4)

$$\bar{r} = \bar{r}_{\mathcal{G}} + \bar{s} \quad \Rightarrow \quad \bar{v} = \bar{v}_{\mathcal{G}} + \dot{\bar{s}} \quad \Leftrightarrow \quad \bar{v} = \bar{v}_{\mathcal{G}} + \bar{\omega} \times \bar{s},$$

where Theorem 12.7 gave that $\dot{\bar{s}} = \bar{\omega} \times \bar{s}$. The arbitrary point in Def. 13.2 is here chosen as $\mathcal{G}$, giving

$$\begin{aligned}
\bar{H}_{\mathcal{G}} &= \int_{\Omega} \bar{s} \times \bar{v} \mathrm{d}m \\
&= \int_{\Omega} \bar{s} \times (\bar{v}_{\mathcal{G}} + \bar{\omega} \times \bar{s}) \, \mathrm{d}m \\
&= \int_{\Omega} \bar{s} \times \bar{v}_{\mathcal{G}} \mathrm{d}m + \int_{\Omega} \bar{s} \times (\bar{\omega} \times \bar{s}) \, \mathrm{d}m = \{\bar{v}_{\mathcal{G}} \text{ const.}\} \\
&= \left(\int_{\Omega} \bar{s} \mathrm{d}m \right) \times \bar{v}_{\mathcal{G}} + \int_{\Omega} \bar{s} \times (\bar{\omega} \times \bar{s}) \, \mathrm{d}m = \{\text{Lemma 13.4}\} \\
&= \int_{\Omega} \bar{s} \times (\bar{\omega} \times \bar{s}) \, \mathrm{d}m. \qquad \square
\end{aligned}$$

Lemma 13.11 asserts that $\bar{H}_{\mathcal{G}}$ is independent of the velocity of the center of mass.

Theorem 13.12 (Angular momentum w.r.t. the center of mass). For a flat rigid body in planar motion, the angular momentum w.r.t. the center of mass $\mathcal{G}$ is

$$\bar{H}_{\mathcal{G}} = I_{\mathcal{G}} \bar{\omega}, \tag{13.19}$$

where $\bar{\omega}$ is the angular velocity of the body, and $I_{\mathcal{G}}$ is the moment of inertia w.r.t. $\mathcal{G}$ (Fig. 13.6).

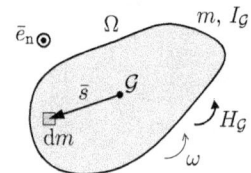

Figure 13.5: The reference plane and the angular momentum for a flat rigid body in planar motion.

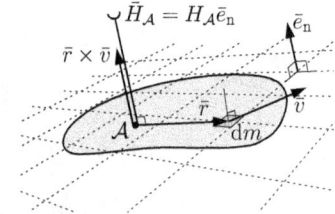

Figure 13.6: Geometry and notation for Theorem 13.12. The angular momentum is $\bar{H}_{\mathcal{G}} = I_{\mathcal{G}} \bar{\omega}$ for a flat rigid body in planar motion.

Proof. For a flat rigid body Ω, Lemma 13.11 yields

$$
\begin{aligned}
\bar{H}_{\mathcal{G}} &= \int_{\Omega} \bar{s} \times (\bar{\omega} \times \bar{s})\, \mathrm{d}m = \big\{\text{Eq. (A.22a)}\big\} \\
&= \int_{\Omega} \big[(\bar{s} \cdot \bar{s})\bar{\omega} - (\bar{s} \cdot \bar{\omega})\, \bar{s}\big]\, \mathrm{d}m = \big\{\bar{s} \perp \bar{\omega} \;\Rightarrow\; \bar{s} \cdot \bar{\omega} = 0\big\} \\
&= \int_{\Omega} (\bar{s} \cdot \bar{s})\bar{\omega}\, \mathrm{d}m = \big\{\bar{\omega} \text{ const.}\big\} \\
&= \bar{\omega} \int_{\Omega} (\bar{s} \cdot \bar{s})\, \mathrm{d}m = \big\{\text{Eq. (13.15)}\big\} \\
&= I_{\mathcal{G}}\bar{\omega}. \qquad\qquad\qquad\qquad\qquad\qquad\qquad \square
\end{aligned}
$$

Thus, the vectors $\bar{H}_{\mathcal{G}}$ and $\bar{\omega}$ have the same direction in planar problems, and this direction is given by the right-hand rule (Fig. 12.5).

Theorem 13.13 (Angular momentum w.r.t. a body- and space-fixed point). For a flat rigid body in planar motion, the angular momentum w.r.t. a body- and space-fixed point $\mathcal{O}$ is

$$
\bar{H}_{\mathcal{O}} = I_{\mathcal{O}}\bar{\omega}, \tag{13.20}
$$

where $\bar{\omega}$ is the angular velocity of the body, and $I_{\mathcal{O}}$ is the moment of inertia of the body w.r.t. $\mathcal{O}$ (Fig. 13.7).

Proof. We place the origin in $\mathcal{O}$ so that $\bar{r}_{\mathcal{G}} = \overline{\mathcal{OG}}$. Since $\bar{r}_{\mathcal{G}}$ is a body-fixed vector, Theorem 12.7 states that

$$
\bar{v}_{\mathcal{G}} = \dot{\bar{r}}_{\mathcal{G}} = \bar{\omega} \times \bar{r}_{\mathcal{G}}. \tag{13.21}
$$

The Transfer theorem of angular momentum for a rigid body, Eq. (13.8), gives

$$
\begin{aligned}
\bar{H}_{\mathcal{O}} &= \bar{H}_{\mathcal{G}} + \overline{\mathcal{OG}} \times m\bar{v}_{\mathcal{G}} = \big\{\text{Theorem 13.12}\big\} \\
&= I_{\mathcal{G}}\bar{\omega} + \bar{r}_{\mathcal{G}} \times m\bar{v}_{\mathcal{G}} = \big\{\text{Eq. (13.21)}\big\} \\
&= I_{\mathcal{G}}\bar{\omega} + m\big[\bar{r}_{\mathcal{G}} \times (\bar{\omega} \times \bar{r}_{\mathcal{G}})\big] = \big\{\text{Eq. (A.22a)}\big\} \\
&= I_{\mathcal{G}}\bar{\omega} + m\,(\bar{r}_{\mathcal{G}} \cdot \bar{r}_{\mathcal{G}})\,\bar{\omega} - m\,(\bar{r}_{\mathcal{G}} \cdot \bar{\omega})\,\bar{r}_{\mathcal{G}} = \big\{\bar{r}_{\mathcal{G}} \perp \bar{\omega}\big\} \\
&= \big(I_{\mathcal{G}} + m|\overline{\mathcal{OG}}|^2\big)\,\bar{\omega} = \big\{\text{Theorem 13.10}\big\} \\
&= I_{\mathcal{O}}\bar{\omega}. \qquad\qquad\qquad\qquad\qquad\qquad\qquad \square
\end{aligned}
$$

For planar problems, the scalar forms of Eqs. (13.19) and (13.20) become

$$
H_{\mathcal{G}} = I_{\mathcal{G}}\omega, \tag{13.22a}
$$

$$
H_{\mathcal{O}} = I_{\mathcal{O}}\omega. \tag{13.22b}
$$

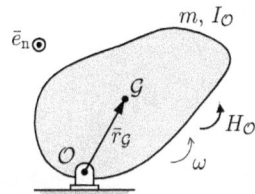

Figure 13.7: Geometry and notation for Theorem 13.13. The angular momentum is $\bar{H}_{\mathcal{O}} = I_{\mathcal{O}}\bar{\omega}$ for a flat rigid body in planar motion w.r.t a body- and space-fixed point $\mathcal{O}$.

Moment equations for planar problems

For planar problems, the moment sum and the angular momentum are both oriented in the normal direction of the reference plane. For this reason, one can write the Moment equation (13.3b) on scalar form:

$$\Sigma M_{\mathcal{D}} = \dot{H}_{\mathcal{D}}, \tag{13.23}$$

where $\mathcal{D}$ is a space-fixed point. In the same way, Eqs. (13.9) and (13.3b) give

$$\Sigma M_{\mathcal{G}} = \dot{H}_{\mathcal{G}}, \tag{13.24a}$$

$$\Sigma M_{\mathcal{O}} = \dot{H}_{\mathcal{O}}, \tag{13.24b}$$

for planar problems, where $\mathcal{G}$ is the center of mass, and $\mathcal{O}$ is a body- and space-fixed point.

In problem solving, the Force equation is typically used together with either Eq. (13.24a) or Eq. (13.24b). The Moment equation w.r.t. $\mathcal{G}$ can always be used (Fig 13.8), giving

$$\begin{aligned}
\Sigma M_{\mathcal{G}} = \dot{H}_{\mathcal{G}} &= \left\{ \text{Eq. (13.22a)} \right\} \\
&= \frac{\mathrm{d}}{\mathrm{d}t}(I_{\mathcal{G}}\omega) = \left\{ I_{\mathcal{G}} \text{ const.} \right\} \\
&= I_{\mathcal{G}}\dot{\omega} \\
&= I_{\mathcal{G}}\alpha.
\end{aligned} \tag{13.25}$$

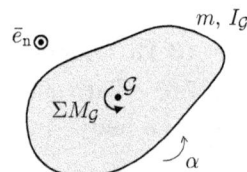

Figure 13.8: The Moment equation, $\Sigma M_{\mathcal{G}} = \dot{H}_{\mathcal{G}} = I_{\mathcal{G}}\alpha$, w.r.t. the center of mass $\mathcal{G}$ of a flat rigid body in planar motion.

If a body- and space-fixed point $\mathcal{O}$ exists (Fig. 13.9), Eq. (13.24b) gives

$$\begin{aligned}
\Sigma M_{\mathcal{O}} = \dot{H}_{\mathcal{O}} &= \left\{ \text{Eq. (13.22b)} \right\} \\
&= \frac{\mathrm{d}}{\mathrm{d}t}(I_{\mathcal{O}}\omega) = \left\{ I_{\mathcal{O}} \text{ const.} \right\} \\
&= I_{\mathcal{O}}\dot{\omega} \\
&= I_{\mathcal{O}}\alpha.
\end{aligned} \tag{13.26}$$

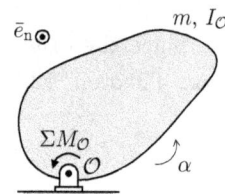

Figure 13.9: The Moment equation, $\Sigma M_{\mathcal{O}} = I_{\mathcal{O}}\alpha$, for a flat rigid body in planar motion about a body- and space-fixed point $\mathcal{O}$.

14
Work–energy method for rigid bodies

14.1 Power and work from forces and couples

According to Def. 9.1, the power of a force $\bar{F}$, whose point of application has the velocity $\bar{v}$, is

$$P \equiv \bar{F} \cdot \bar{v}.$$

This power of a force is used to define the work of a force between times t_1 and t_2 (Def. 9.2):

$$U_{1-2} \equiv \int_{t_1}^{t_2} P \mathrm{d}t = \int_{t_1}^{t_2} \bar{F} \cdot \bar{v} \mathrm{d}t = \int_{s_1}^{s_2} \bar{F} \cdot \bar{e}_t \mathrm{d}s,$$

where the latter integral represents work between two positions on the path of the point of application of the force (Theorem 9.3).

Besides forces, couples may act on a rigid body. The power of a couple is the power from the implicit force pair that creates this couple.

Theorem 14.1 (Power of a couple). The power of a couple $\bar{C}$, acting on a rigid body in planar motion with angular velocity $\bar{\omega}$, is

$$P = \bar{C} \cdot \bar{\omega}. \tag{14.1}$$

Proof. Let the couple $\bar{C}$ be represented by a force pair, $\bar{F}$ and $-\bar{F}$ acting in the body-fixed points $\mathcal{P}$ and $\mathcal{Q}$, respectively (Fig. 14.1). This force pair produces the power (Def. 9.1)

$$
\begin{aligned}
P &= \bar{F} \cdot \bar{v}_{\mathcal{P}} + (-\bar{F}) \cdot \bar{v}_{\mathcal{Q}} = \{\text{Eq. (12.6)}\} \\
&= \bar{F} \cdot \bar{v}_{\mathcal{P}} - \bar{F} \cdot (\bar{v}_{\mathcal{P}} + \bar{\omega} \times \overline{\mathcal{PQ}}) = \{\overline{\mathcal{QP}} = -\overline{\mathcal{PQ}}\} \\
&= \bar{F} \cdot (\bar{\omega} \times \overline{\mathcal{QP}}) = \{\text{Eq. (A.22b)}\} \\
&= \bar{\omega} \cdot (\overline{\mathcal{QP}} \times \bar{F}) = \{\text{Theorem 2.6}\} \\
&= \bar{\omega} \cdot \bar{C}. \qquad \square
\end{aligned}
$$

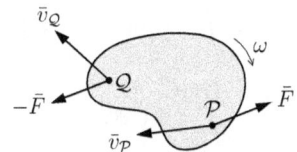

Figure 14.1: Geometry and notation for Theorem 14.1.

For a flat rigid body in planar motion, we have that $\bar{\omega} = \omega\bar{e}_n$ and $\bar{C} = C\bar{e}_n$, where $\bar{e}_n$ is the normal of the reference plane. Thus, the power of the couple can be written as

$$P = (C\bar{e}_n) \cdot (\omega\bar{e}_n) = C\omega. \tag{14.2}$$

The work of a couple is defined as the time integral of the power of the couple. This definition is analogous to that of the work of a force.

Definition 14.2 (Work of a couple). The work of a couple $\bar{C}$, acting on a flat rigid body in planar motion between times t_1 and t_2, is

$$U_{1-2} \equiv \int_{t_1}^{t_2} P dt = \int_{t_1}^{t_2} \bar{C} \cdot \bar{\omega} dt, \tag{14.3}$$

where $P = \bar{C} \cdot \bar{\omega}$ is the power of the couple, and $\bar{\omega}$ is the angular velocity of the body.

The integral over time in Eq. (14.3) can be rewritten as an integral over orientation.

Theorem 14.3 (Work between orientations). The work of a couple $\bar{C} = C\bar{e}_n$, that acts on a rigid body in planar motion between orientations 1 and 2, is

$$U_{1-2} = \int_{\theta_1}^{\theta_2} C d\theta, \tag{14.4}$$

where $\bar{e}_n$ is the normal of the reference plane, θ is the polar angle of a body-fixed axis, and θ_1 and θ_2 denote the polar angle of orientations 1 and 2, respectively (Fig. 14.2).

Proof. Let t_1 and t_2 be the times corresponding to orientations 1 and 2. According to Def. 12.4, $\bar{\omega} = \dot{\theta}\bar{e}_n$, so that $\bar{C} \cdot \bar{\omega} = C\dot{\theta}$. Substitution into Eq. (14.3) gives

$$U_{1-2} = \int_{t_1}^{t_2} C\dot{\theta} dt$$
$$= \int_{t_1}^{t_2} C[\theta(t)]\frac{d\theta}{dt} dt = \left\{ \begin{matrix} \text{subst. according to (A.39)} \\ \theta = \theta(t),\, d\theta = \frac{d\theta}{dt} dt \end{matrix} \right\}$$
$$= \int_{\theta(t_1)}^{\theta(t_2)} C d\theta. \qquad \square$$

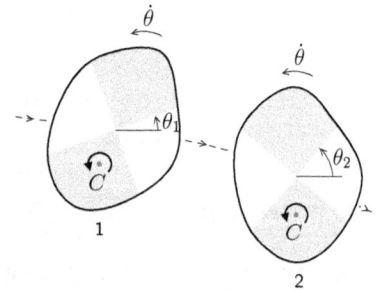

Figure 14.2: Geometry and notation for Theorem 14.3.

14.2 Power sum and work on a rigid body

Definition 14.4 (Power sum). The *power sum* ΣP of a rigid body is the sum of the powers from all forces and couples acting on the rigid body.

Consider a flat rigid body in planar motion which is subjected to n forces and m couples. According to Def. 14.4, using the notation from Fig. 14.3, the power sum is

$$\Sigma P = \sum_{i=1}^{n} \bar{F}_i \cdot \bar{v}_i + \sum_{j=1}^{m} \bar{C}_j \cdot \bar{\omega}. \tag{14.5}$$

This sum includes the power from every force (Def. 9.1) and every couple (Theorem 14.1) of the force-couple system.

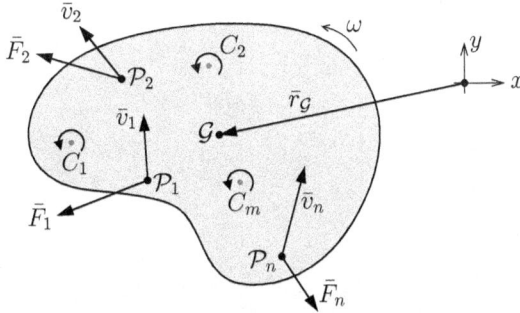

Figure 14.3: Geometry and notation for the definition of the power sum.

Theorem 14.5. The power sum of a force-couple system, acting on a flat rigid body in planar motion, is

$$\Sigma P = \Sigma \bar{F} \cdot \bar{v}_{\mathcal{G}} + \Sigma \bar{M}_{\mathcal{G}} \cdot \bar{\omega}, \tag{14.6}$$

where $\bar{v}_{\mathcal{G}}$ is the velocity of the center of mass $\mathcal{G}$, $\bar{\omega}$ is the angular velocity of the body, $\Sigma \bar{F}$ is the power sum, and $\Sigma \bar{M}_{\mathcal{G}}$ is the moment sum w.r.t. $\mathcal{G}$ (Fig. 14.4).

Proof. Using the notation in Fig. 14.3, we let $\bar{r}_i$ be the position vector of $\mathcal{P}_i$, and let $\bar{s}_i = \overline{\mathcal{G}\mathcal{P}_i}$. Then, the Parallelogram law gives

$$\bar{r}_i = \bar{r}_{\mathcal{G}} + \bar{s}_i \quad \Rightarrow \quad \bar{v}_i = \bar{v}_{\mathcal{G}} + \dot{\bar{s}}_i \quad \Leftrightarrow \quad \bar{v}_i = \bar{v}_{\mathcal{G}} + \bar{\omega} \times \bar{s}_i,$$

where Theorem 12.7 gave that $\dot{\bar{s}}_i = \bar{\omega} \times \bar{s}_i$. According to Eq. (14.5), the power sum is

$$\begin{aligned}
\Sigma P &= \sum_{i=1}^{n} \bar{F}_i \cdot \bar{v}_i + \sum_{j=1}^{m} \bar{C}_j \cdot \bar{\omega} \\
&= \sum_{i=1}^{n} \bar{F}_i \cdot (\bar{v}_{\mathcal{G}} + \bar{\omega} \times \bar{s}_i) + \sum_{i=j}^{m} \bar{C}_j \cdot \bar{\omega} = \{\text{Def. 2.8}\} \\
&= \Sigma \bar{F} \cdot \bar{v}_{\mathcal{G}} + \sum_{i=1}^{n} \bar{F}_i \cdot (\bar{\omega} \times \bar{s}_i) + \sum_{j=1}^{m} \bar{C}_j \cdot \bar{\omega} = \{\text{Eq. (A.22b)}\} \\
&= \Sigma \bar{F} \cdot \bar{v}_{\mathcal{G}} + \sum_{i=1}^{n} \bar{\omega} \cdot (\bar{s}_i \times \bar{F}_i) + \sum_{j=1}^{m} \bar{C}_j \cdot \bar{\omega}
\end{aligned}$$

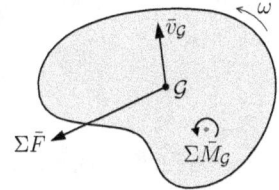

Figure 14.4: Force and moment sum w.r.t. the center of mass $\mathcal{G}$ for the system in Fig. 14.3.

$$= \Sigma \bar{F} \cdot \bar{v}_{\mathcal{G}} + \left[\sum_{i=1}^{n} \overline{\mathcal{GP}_i} \times \bar{F}_i + \sum_{j=1}^{m} \bar{C}_j \right] \cdot \bar{\omega} = \{\text{Def. 2.9}\}$$

$$= \Sigma \bar{F} \cdot \bar{v}_{\mathcal{G}} + \Sigma \bar{M}_{\mathcal{G}} \cdot \bar{\omega}. \qquad \qquad \square$$

Definition 14.6 (Work on a rigid body). The work ΣU_{1-2} on a rigid body in planar motion between times t_1 and t_2 is

$$\Sigma U_{1-2} \equiv \int_{t_1}^{t_2} \Sigma P \mathrm{d}t, \qquad \qquad (14.7)$$

where ΣP is the power sum for the rigid body.

By substituting Eq. (14.5) into Eq. (14.7), we obtain

$$\Sigma U_{1-2} = \int_{t_1}^{t_2} \left(\sum_{i=1}^{n} \bar{F}_i \cdot \bar{v}_i + \sum_{j=1}^{m} \bar{C}_j \cdot \bar{\omega} \right) \mathrm{d}t$$

$$= \sum_{i=1}^{n} \int_{t_1}^{t_2} \bar{F}_i \cdot \bar{v}_i \mathrm{d}t + \sum_{j=1}^{m} \int_{t_1}^{t_2} \bar{C}_j \cdot \bar{\omega} \mathrm{d}t, \qquad (14.8)$$

which gives us the insight that the work on a rigid body is the sum of the work from every force and couple that act on this rigid body.

14.3   Kinetic energy

Definition 14.7 (Kinetic energy). The *kinetic energy* of a body Ω is defined as

$$K \equiv \frac{1}{2} \int_{\Omega} (\bar{v} \cdot \bar{v}) \mathrm{d}m = \frac{1}{2} \int_{\Omega} v^2 \mathrm{d}m, \qquad \qquad (14.9)$$

where $\bar{v}$ denotes the velocity of a mass element $\mathrm{d}m$ in an inertial system (Fig. 13.1).

Consequently, the kinetic energy of a body is the sum of the contributions $\mathrm{d}K = \frac{1}{2}v^2 \mathrm{d}m$ from its mass elements.

Theorem 14.8 (Kinetic energy of a flat rigid body). The kinetic energy of a flat rigid body in planar motion is

$$K = \frac{1}{2}mv_{\mathcal{G}}^2 + \frac{1}{2}I_{\mathcal{G}}\omega^2, \qquad \qquad (14.10)$$

where m is the mass of the body, $I_{\mathcal{G}}$ is its moment of inertia w.r.t. the center of mass $\mathcal{G}$, $v_{\mathcal{G}}$ is the speed of $\mathcal{G}$, and ω is the angular velocity of the body.

Proof. Let $\bar{r}_{\mathcal{G}}$ be the position of $\mathcal{G}$, so that the position vector $\bar{r}$ of the mass element can be written as (Fig. 13.4)

$$\bar{r} = \bar{r}_{\mathcal{G}} + \bar{s} \quad \Rightarrow \quad \bar{v} = \bar{v}_{\mathcal{G}} + \dot{\bar{s}} \quad \Leftrightarrow \quad \bar{v} = \bar{v}_{\mathcal{G}} + \bar{\omega} \times \bar{s},$$

where Theorem 12.7 gave that $\dot{\bar{s}} = \bar{\omega} \times \bar{s}$. Using Def. 14.7 for the rigid body Ω gives

$$
\begin{aligned}
K &= \frac{1}{2} \int_{\Omega} (\bar{v} \cdot \bar{v}) \mathrm{d}m \\
&= \frac{1}{2} \int_{\Omega} (\bar{v}_{\mathcal{G}} + \bar{\omega} \times \bar{s}) \cdot (\bar{v}_{\mathcal{G}} + \bar{\omega} \times \bar{s}) \, \mathrm{d}m \\
&= \frac{1}{2} \int_{\Omega} \left[|\bar{v}_{\mathcal{G}}|^2 + 2\bar{v}_{\mathcal{G}} \cdot (\bar{\omega} \times \bar{s}) + |\bar{\omega} \times \bar{s}|^2 \right] \mathrm{d}m = \{\bar{\omega}, \bar{v}_{\mathcal{G}} \text{ const.}\} \\
&= \frac{1}{2} v_{\mathcal{G}}^2 \underbrace{\int_{\Omega} \mathrm{d}m}_{=m} + \bar{v}_{\mathcal{G}} \cdot \left(\bar{\omega} \times \underbrace{\int_{\Omega} \bar{s}\,\mathrm{d}m}_{=\bar{0}} \right) + \frac{1}{2} \int_{\Omega} |\bar{\omega} \times \bar{s}|^2 \mathrm{d}m = \{\bar{\omega} \perp \bar{s}\} \\
&= \frac{1}{2} m v_{\mathcal{G}}^2 + \frac{1}{2} \int_{\Omega} |\bar{\omega}|^2 |\bar{s}|^2 \mathrm{d}m = \{\bar{\omega} \text{ const.}\} \\
&= \frac{1}{2} m v_{\mathcal{G}}^2 + \frac{1}{2} \omega^2 \int_{\Omega} (\bar{s} \cdot \bar{s}) \mathrm{d}m = \{\text{Eq. (13.15)}\} \\
&= \frac{1}{2} m v_{\mathcal{G}}^2 + \frac{1}{2} I_{\mathcal{G}} \omega^2. \qquad \qquad \square
\end{aligned}
$$

Theorem 14.9 (Kinetic energy for fixed-axis rotation). The kinetic energy of a flat rigid body rotating about a body- and space-fixed point $\mathcal{O}$ is

$$
K = \frac{1}{2} I_{\mathcal{O}} \omega^2, \tag{14.11}
$$

where $I_{\mathcal{O}}$ is the moment of inertia w.r.t. $\mathcal{O}$, and ω is the angular velocity of the body.

Proof. Since $\mathcal{O}$ is the instantaneous center of the rigid body, it holds that

$$
v_{\mathcal{G}} = \pm |\overline{\mathcal{O}\mathcal{G}}| \omega \qquad \Leftrightarrow \qquad v_{\mathcal{G}}^2 = |\overline{\mathcal{O}\mathcal{G}}|^2 \omega^2. \tag{14.12}
$$

Then, Eq. (14.10) gives

$$
\begin{aligned}
K &= \frac{1}{2} m v_{\mathcal{G}}^2 + \frac{1}{2} I_{\mathcal{G}} \omega^2 = \{\text{Eq. (14.12)}\} \\
&= \frac{1}{2} m |\overline{\mathcal{O}\mathcal{G}}|^2 \omega^2 + \frac{1}{2} I_{\mathcal{G}} \omega^2 \\
&= \frac{1}{2} \left(m |\overline{\mathcal{O}\mathcal{G}}|^2 + I_{\mathcal{G}} \right) \omega^2 = \{\text{Theorem 13.10}\} \\
&= \frac{1}{2} I_{\mathcal{O}} \omega^2. \qquad \qquad \square
\end{aligned}
$$

14.4 Work–energy theorem

The work applied to the rigid body by the force-couple system is transferred into kinetic energy. This process is described by the *Work–energy theorem* for rigid bodies.

Theorem 14.10 (Work–energy theorem). For a flat rigid body in planar motion, subjected to an arbitrary force-couple system between positions 1 and 2 (Fig. 2.7), it holds that

$$\Sigma U_{1-2} = K_2 - K_1, \tag{14.13}$$

where ΣU_{1-2} is the work applied to the body, while K_1 and K_2 are the kinetic energies of the body in positions 1 and 2, respectively.

Proof. Time differentiation of Eq. (14.10) gives

$$\frac{dK}{dt} = \frac{d}{dt}\left[\frac{1}{2}m(\bar{v}_{\mathcal{G}} \cdot \bar{v}_{\mathcal{G}}) + \frac{1}{2}I_{\mathcal{G}}(\bar{\omega} \cdot \bar{\omega})\right] = \{\text{Product rule (A.25b)}\}$$

$$= \frac{1}{2}m(\bar{a}_{\mathcal{G}} \cdot \bar{v}_{\mathcal{G}}) + \frac{1}{2}m(\bar{v}_{\mathcal{G}} \cdot \bar{a}_{\mathcal{G}}) + \frac{1}{2}I_{\mathcal{G}}(\bar{\alpha} \cdot \bar{\omega}) + \frac{1}{2}I_{\mathcal{G}}(\bar{\omega} \cdot \bar{\alpha})$$

$$= m\bar{a}_{\mathcal{G}} \cdot \bar{v}_{\mathcal{G}} + I_{\mathcal{G}}\bar{\alpha} \cdot \bar{\omega} = \{\text{Force and Moment equations}\}$$

$$= \Sigma\bar{F} \cdot \bar{v}_{\mathcal{G}} + \Sigma\bar{M}_{\mathcal{G}} \cdot \bar{\omega} = \{\text{Theorem 14.5}\}$$

$$= \Sigma P,$$

where ΣP is the power sum of the body. This relation can, according to Eq. (A.30), be expressed using differential notation:

$$\Sigma P dt = dK \quad \Leftrightarrow \quad \{\text{Theorem A.3}\} \quad \Leftrightarrow$$

$$\int_{t_1}^{t_2} \Sigma P dt = \int_{K_1}^{K_2} dK \quad \Leftrightarrow \quad \{\text{Def. 14.6}\} \quad \Leftrightarrow$$

$$\Sigma U_{1-2} = K_2 - K_1,$$

where t_1 and t_2 are the times corresponding to positions 1 and 2. $\square$

In the same way as for particles, we can identify conservative forces for rigid bodies. These forces preserve the total mechanical energy while carrying out work. We already know from Theorem 9.10 that the work of a spring force can be expressed as

$$U_{1-2} = -\left[V_e(\ell_2) - V_e(\ell_1)\right], \tag{14.14}$$

where $V_e(\ell) = \frac{1}{2}k(\ell - \ell_0)^2$ is the elastic energy. To express the work of the force of gravity, it is necessary to define the potential energy of a rigid body.

Definition 14.11 (Potential energy of a rigid body). The *potential energy* of a rigid body, with mass m in a uniform field of gravity $\bar{g} = -g\bar{e}_y$, is

$$V_g(y_{\mathcal{G}}) \equiv mgy_{\mathcal{G}}, \tag{14.15}$$

where $y_{\mathcal{G}}$ is the height coordinate of the center of mass $\mathcal{G}$ of the body relative to an inertial system.

The work of the force of gravity on a rigid body can be determined in the same way as the work of the force of gravity on a particle.

Theorem 14.12 (Work of the force of gravity on a rigid body). For a rigid body with mass m in a uniform field of gravity $\bar{g} = -g\bar{e}_y$, the work of the force of gravity $m\bar{g}$ between positions 1 and 2 is

$$U_{1-2} = -\left[V_\mathrm{g}(y_{\mathcal{G}2}) - V_\mathrm{g}(y_{\mathcal{G}1})\right], \tag{14.16}$$

where $y_{\mathcal{G}1}$ and $y_{\mathcal{G}2}$ are the height coordinates of the center of mass $\mathcal{G}$ of the rigid body in positions 1 and 2, respectively, and $V_\mathrm{g}(y_\mathcal{G})$ is the potential energy of the rigid body (Fig. 14.5).

Proof. Theorem 14.12 is proved in the same way as Theorem 9.8. ☐

The Work–energy theorem for rigid bodies can now be rewritten with potential terms in the same way as the Work–energy theorem for particles:

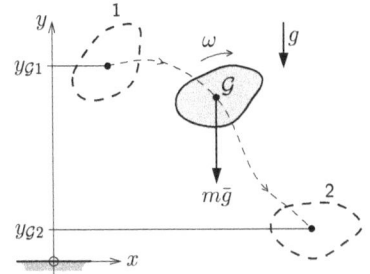

Figure 14.5: Geometry for the work of the force of gravity on a rigid body.

$$\Sigma U_{1-2} = K_2 - K_1 \quad \Leftrightarrow$$
$$-(V_\mathrm{g2} - V_\mathrm{g1}) - (V_\mathrm{e2} - V_\mathrm{e1}) + \Sigma U'_{1-2} = K_2 - K_1,$$

where we use Theorems 14.12 and 9.10, and let $\Sigma U'_{1-2}$ denote the work of all forces and all couples, excluding work from the force of gravity and spring forces. Hence, we have

$$\Sigma U'_{1-2} = (V_\mathrm{g2} - V_\mathrm{g1}) + (V_\mathrm{e2} - V_\mathrm{e1}) + (K_2 - K_1). \tag{14.17}$$

15

Impulse relations for rigid bodies

If the forces and couples that act on a rigid body are time-dependent, but cannot be written as functions of the position and orientation of the rigid body in any obvious way, then the Work–energy method becomes difficult to employ. As an alternative, impulse relations can be useful.

15.1 Integral form of Euler's laws

Euler's laws of motion, Postulate 13.3, are formulated for general, deformable bodies. By integrating these laws w.r.t. time, we obtain impulse relations for general bodies.

Theorem 15.1 (Impulse–momentum relation). If a body is affected by a force-couple system with the force sum $\Sigma\bar{F}$ between times t_1 and t_2, then it holds that

$$\int_{t_1}^{t_2} \Sigma\bar{F}\mathrm{d}t = \bar{G}(t_2) - \bar{G}(t_1), \qquad (15.1)$$

where $\bar{G}$ is the momentum of the body.

Proof. Euler's first law, Eq. (13.3a), gives

$$\Sigma\bar{F} = \frac{\mathrm{d}\bar{G}}{\mathrm{d}t} \quad \Leftrightarrow \quad \big\{\text{Eq. (A.35)}\big\} \quad \Leftrightarrow$$

$$\Sigma\bar{F}\mathrm{d}t = d\bar{G} \quad \Leftrightarrow \quad \big\{\text{Eq. (A.36)}\big\} \quad \Leftrightarrow$$

$$\int_{t_1}^{t_2} \Sigma\bar{F}\mathrm{d}t = \bar{G}(t_2) - \bar{G}(t_1). \qquad \square$$

According to Def. 10.3, the integral in Eq. (15.1) is the impulse of the force sum. By integrating the Moment equation (13.3b) w.r.t. time, we obtain the *Angular impulse–angular momentum relation*.

Theorem 15.2 (Angular impulse–angular momentum relation). If a body is affected by a force-couple system with the moment sum

$\Sigma \bar{M}_{\mathcal{D}}$ w.r.t. a space-fixed point $\mathcal{D}$ between times t_1 and t_2, then it holds that

$$\int_{t_1}^{t_2} \Sigma \bar{M}_{\mathcal{D}} \mathrm{d}t = \bar{H}_{\mathcal{D}}(t_2) - \bar{H}_{\mathcal{D}}(t_1), \tag{15.2}$$

where $\bar{H}_{\mathcal{D}}$ is the angular momentum of the body w.r.t. $\mathcal{D}$.

Proof. Euler's second law, Eq. (13.3b), gives

$$\Sigma \bar{M}_{\mathcal{D}} = \frac{\mathrm{d}\bar{H}_{\mathcal{D}}}{\mathrm{d}t} \quad \Leftrightarrow \quad \{\text{Eq. (A.35)}\} \quad \Leftrightarrow$$

$$\Sigma \bar{M}_{\mathcal{D}} \mathrm{d}t = \mathrm{d}\bar{H}_{\mathcal{D}} \quad \Leftrightarrow \quad \{\text{Eq. (A.36)}\} \quad \Leftrightarrow$$

$$\int_{t_1}^{t_2} \Sigma \bar{M}_{\mathcal{D}} \mathrm{d}t = \bar{H}_{\mathcal{D}}(t_2) - \bar{H}_{\mathcal{D}}(t_1). \qquad \square$$

The time integral on the left-hand side of Eq. (15.2) is called the *angular impulse* of the moment sum w.r.t. $\mathcal{D}$. With notation as in Def. 2.7, this angular impulse consists of contributions from the moments of force and couples of the force-couple system:

$$\int_{t_1}^{t_2} \Sigma \bar{M}_{\mathcal{D}} \mathrm{d}t = \sum_{i=1}^{n} \int_{t_1}^{t_2} \overline{\mathcal{DP}_i} \times \bar{F}_i \mathrm{d}t + \sum_{j=1}^{m} \int_{t_1}^{t_2} \bar{C}_j \mathrm{d}t.$$

Note that the points of application $\mathcal{P}_i$ of the forces move, which implies that the vectors $\overline{\mathcal{DP}_i}$ are time-dependent. The contribution of a couple to the angular impulse is called an *impulse couple*.

Definition 15.3 (Impulse couple). The *impulse couple* of a couple $\bar{C}$, that acts between times t_1 and t_2, is

$$\bar{J} \equiv \int_{t_1}^{t_2} \bar{C} \mathrm{d}t. \tag{15.3}$$

An impulse couple $\bar{J}$ can also be created by an *impulse pair*, *i.e.* two oppositely directed impulses of equal magnitude (Fig. 15.1).

15.2 Impulse relations for rigid bodies

The Impulse–momentum relation and the Angular impulse–angular momentum relation derived above are both formulated for deformable bodies. In the case of rigid bodies, it holds that $\bar{G} = mv_{\mathcal{G}}$, so that the Impulse–momentum relation, Eq. (15.1), can be written

$$\int_{t_1}^{t_2} \Sigma \bar{F} \mathrm{d}t = m\bar{v}_{\mathcal{G}}(t_2) - m\bar{v}_{\mathcal{G}}(t_1), \tag{15.4}$$

where $\mathcal{G}$ is the center of mass. Furthermore, if we integrate the Moment equation for rigid bodies w.r.t. $\mathcal{G}$, Eq. (13.9), we obtain a new variant of the Angular impulse–angular momentum relation.

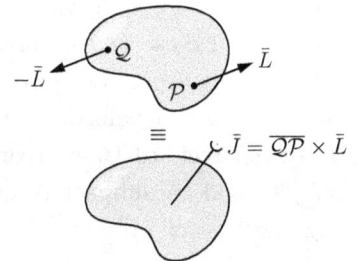

Figure 15.1: The impulse couple of two equally-sized and oppositely directed impulses.

Theorem 15.4 (Angular impulse–angular momentum relation w.r.t. the center of mass). If a rigid body is affected by a force-couple system, with the moment sum $\Sigma \bar{M}_{\mathcal{G}}$ w.r.t. the center of mass $\mathcal{G}$ between times t_1 and t_2, it holds that

$$\int_{t_1}^{t_2} \Sigma \bar{M}_{\mathcal{G}} dt = \bar{H}_{\mathcal{G}}(t_2) - \bar{H}_{\mathcal{G}}(t_1), \tag{15.5}$$

where $\bar{H}_{\mathcal{G}}$ is the angular momentum of the rigid body w.r.t. $\mathcal{G}$.

Proof. The Moment equation w.r.t. $\mathcal{G}$, Eq. (13.9), gives

$$\Sigma \bar{M}_{\mathcal{G}} = \frac{d\bar{H}_{\mathcal{G}}}{dt} \quad \Leftrightarrow \quad \{\text{Eq. (A.35)}\} \quad \Leftrightarrow$$

$$\Sigma \bar{M}_{\mathcal{G}} dt = d\bar{H}_{\mathcal{G}} \quad \Leftrightarrow \quad \{\text{Eq. (A.36)}\} \quad \Leftrightarrow$$

$$\int_{t_1}^{t_2} \Sigma \bar{M}_{\mathcal{G}} dt = \bar{H}_{\mathcal{G}}(t_2) - \bar{H}_{\mathcal{G}}(t_1). \qquad \square$$

15.3   Impact with rigid bodies

Consider a rigid body which collides with another body. As a result, the velocity and the angular velocity of the rigid body will change considerably during a short interval of time. According to Euler's laws of motion, this implies that the force and moment sums that act on the rigid body grow very large during the collision. Some forces that act on the rigid body, for instance the force of gravity, vary within bounds during the impact process. In contrast, contact forces and couples grow without bound; we call these *impact forces* and *impact couples*, respectively.

Let us study an impact process beginning at time $t = 0$ and ending at time $t = \Delta t$. Analogously to the instantaneous impact model for particles (*cf.* Sect. 10.4), we assume that the impact forces, $\bar{F}_1^{\text{imp}}$, $\bar{F}_2^{\text{imp}}$, ..., $\bar{F}_n^{\text{imp}}$, and the impact couples, $\bar{C}_1^{\text{imp}}$, $\bar{C}_2^{\text{imp}}$, ..., $\bar{C}_m^{\text{imp}}$, dominate during the impact process. Thus, the impact forces are integrated to impact impulses

$$\bar{L}_i^{\text{imp}} = \int_0^{\Delta t} \bar{F}_i^{\text{imp}} dt, \qquad i = 1, \dots, n,$$

and the impact couples are integrated to impact impulse couples

$$\bar{J}_i^{\text{imp}} = \int_0^{\Delta t} \bar{C}_i^{\text{imp}} dt, \qquad i = 1, \dots, m,$$

while other impulses and impulse couples are neglected. Similarly to the concept of the force-couple system, the impulses and impulse couples of the instantaneous impact model form an *impulse system* (Fig. 15.2).

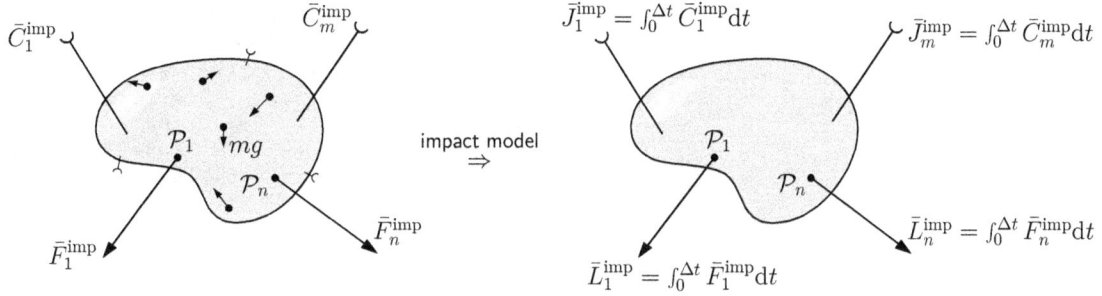

Figure 15.2: A force-couple system, where impact forces $\bar{F}_i^{\text{imp}}$ and impact couples $\bar{C}_j^{\text{imp}}$ dominate. By using an instantaneous impact model, an impulse system is formed with impulses $\bar{L}_i^{\text{imp}}$ and impact couples $\bar{J}_j^{\text{imp}}$.

Let a space-fixed point $\mathcal{D}$ be the reference point during an impact process. The momentum and the angular momentum of the rigid body, before the impact, are denoted by $\bar{G}$ and $\bar{H}_\mathcal{D}$, respectively, whereas $\bar{G}'$ and $\bar{H}'_\mathcal{D}$ denote the same quantities after the impact. We use these quantities to express Eqs. (15.1) and (15.2) for the impact process of the rigid body:

$$\sum_{i=1}^{n} \bar{L}_i^{\text{imp}} = \bar{G}' - \bar{G}, \qquad (15.6a)$$

$$\sum_{i=1}^{n} \overline{\mathcal{DP}_i} \times \bar{L}_i^{\text{imp}} + \sum_{j=1}^{m} \bar{J}_j^{\text{imp}} = \bar{H}'_\mathcal{D} - \bar{H}_\mathcal{D}. \qquad (15.6b)$$

We obtained Eq. (15.6b) by approximating the angular impulse as,

$$\int_0^{\Delta t} \overline{\mathcal{DP}_i} \times \bar{F}_i^{\text{imp}} dt \approx \overline{\mathcal{DP}_i} \times \int_0^{\Delta t} \bar{F}_i^{\text{imp}} dt = \overline{\mathcal{DP}_i} \times \bar{L}_i^{\text{imp}}, \qquad i = 1, \ldots, n,$$

which is motivated by an assumption that the points $\mathcal{P}_i$ move a negligible distance during the instantaneous impact. We illustrate the use of Eqs. (15.6a) and (15.6b) with two examples.

A door with mass m and width b has the counterclockwise angular velocity ω when it closes and locks (Fig. 15.3a). We draw a free-body diagram with the impulse components $L_{\mathcal{O}x}^{\text{imp}}$, $L_{\mathcal{O}y}^{\text{imp}}$ and $L_{\mathcal{A}}^{\text{imp}}$ (Fig. 15.3b). Thereafter, we write the Impulse–momentum relation and the Angular impulse–angular momentum relation, Eqs. (15.6a) and (15.6b), on component form for this planar impact problem:

$$\to^x: \qquad L_{\mathcal{O}x}^{\text{imp}} = 0 - 0,$$

$$\uparrow^y: \qquad L_{\mathcal{O}y}^{\text{imp}} + L_{\mathcal{A}}^{\text{imp}} = 0 - \left(-\frac{1}{2}mb\omega\right),$$

$$\stackrel{\frown}{\mathcal{O}}: \qquad -L_{\mathcal{A}}^{\text{imp}}b = 0 - \frac{1}{3}mb^2\omega,$$

where we used that $I_\mathcal{O} = \frac{1}{3}mb^2$ for the door.[26]

Next, we study a slender bar, with mass m and length ℓ, rotating in the vertical plane about a hinge $\mathcal{O}$. When the bar forms an angle γ

[26] To calculate $I_\mathcal{O}$, use Table C.2 with Steiner's theorem 13.10.

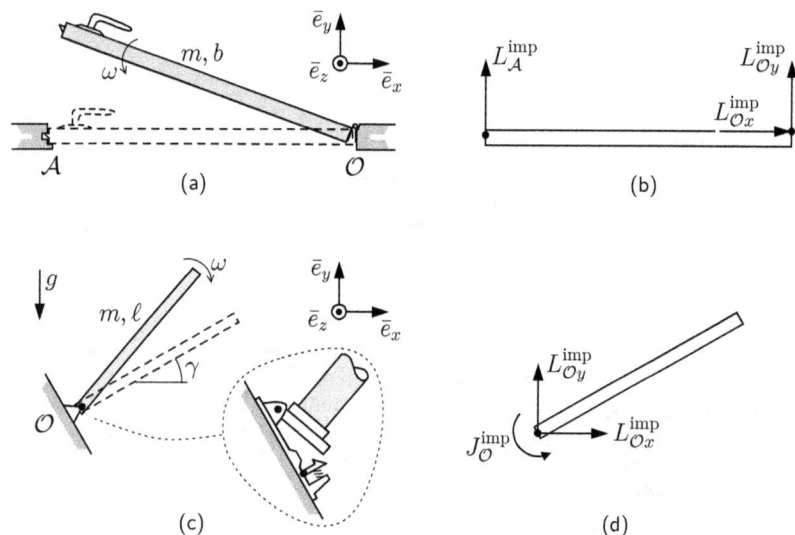

Figure 15.3: (a) A door closes and locks. (b) A free-body diagram of the door with the impulse system for the locking process. (c) A bar is stopped by a mechanism at the hinge $\mathcal{O}$. (d) A free-body diagram of the bar with the impulse system for the impact process.

with a horizontal plane, at the clockwise angular velocity ω, the rotation is stopped by a mechanism that secures the bar (Fig. 15.3c). We draw a free-body diagram, including the impulse components $L_{\mathcal{O}x}^{\mathrm{imp}}$ and $L_{\mathcal{O}y}^{\mathrm{imp}}$, and an impulse couple $J_{\mathcal{O}}^{\mathrm{imp}}$. The latter appears because rotation is prohibited by the mechanism at $\mathcal{O}$ (Fig. 15.3d). Equations (15.6a) and (15.6b) give

$$\rightarrow^x: \qquad L_{\mathcal{O}x}^{\mathrm{imp}} = 0 - \frac{1}{2}m\ell\omega\sin\gamma,$$

$$\uparrow^y: \qquad L_{\mathcal{O}y}^{\mathrm{imp}} = 0 - \left(-\frac{1}{2}m\ell\omega\cos\gamma\right),$$

$$\stackrel{\frown}{\mathcal{O}}: \qquad J_{\mathcal{O}}^{\mathrm{imp}} = 0 - \left(-\frac{1}{3}m\ell^2\omega\right),$$

where we used that $I_{\mathcal{O}} = \frac{1}{3}m\ell^2$ for a slender bar. Note that the impulse from gravity,

$$\bar{L}_{\mathrm{g}} = \int_0^{\Delta t} m\bar{g}\mathrm{d}t = m\bar{g}\Delta t,$$

for the duration Δt of the impact, is neglected; for an instantaneous impact, when $\Delta t \to 0$, the impulse from the force of gravity vanishes.

If an impulse problem involves several rigid bodies, free-body diagrams of their respective impulse systems are drawn, and the Impulse–momentum relation and Angular impulse–angular momentum relation are formulated for each body.

16

Three-dimensional kinematics of rigid bodies

To extend the theory of rigid bodies from planar motion to general, three-dimensional motion, it is necessary to redefine several concepts, including the angular velocity $\bar{\omega}$ and the angular acceleration $\bar{\alpha}$ of rigid bodies.

16.1 Angular velocity and angular acceleration

Similarly to planar motion, we consider a space-fixed, orthogonal coordinate system XYZ with a constant basis $\{\bar{e}_X, \bar{e}_Y, \bar{e}_Z\}$. In addition, we introduce a second coordinate system xyz with the basis $\{\bar{e}_x, \bar{e}_y, \bar{e}_z\}$ (Fig. 16.2). This latter coordinate system xyz is allowed to move and reorient in space, while its orthogonality is preserved.

Definition 16.1 (Angular velocity). The *angular velocity* of a coordinate system xyz in general, three-dimensional motion is

$$\bar{\Omega} \equiv (\dot{\bar{e}}_y \cdot \bar{e}_z)\bar{e}_x + (\dot{\bar{e}}_z \cdot \bar{e}_x)\bar{e}_y + (\dot{\bar{e}}_x \cdot \bar{e}_y)\bar{e}_z, \tag{16.1}$$

where $\{\bar{e}_x, \bar{e}_y, \bar{e}_z\}$ is the basis of xyz.

The basis vectors, $\bar{e}_x$, $\bar{e}_y$ and $\bar{e}_z$, are always perpendicular to each other, even though their directions vary with time. We define the angular velocity of a rigid body by identifying it with the angular velocity of a body-fixed coordinate system xyz (Fig. 16.2).

Definition 16.2 (Angular velocity of a rigid body). The angular velocity of a rigid body in general, three-dimensional motion is

$$\bar{\omega} \equiv \bar{\Omega}, \tag{16.2}$$

where $\bar{\Omega} = (\dot{\bar{e}}_y \cdot \bar{e}_z)\bar{e}_x + (\dot{\bar{e}}_z \cdot \bar{e}_x)\bar{e}_y + (\dot{\bar{e}}_x \cdot \bar{e}_y)\bar{e}_z$ is the angular velocity of a body-fixed coordinate system xyz (Fig. 16.2).

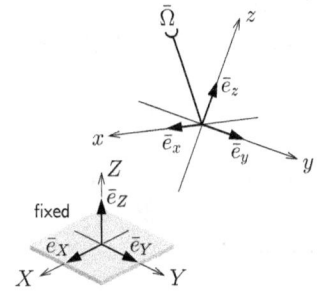

Figure 16.1: A coordinate system xyz with time-dependent coordinate directions, rotating with the angular velocity $\bar{\Omega}$ relative to a space-fixed coordinate system XYZ.

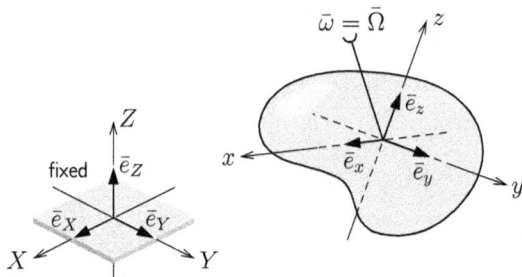

Figure 16.2: The coordinate system xyz is body-fixed, and its angular velocity $\bar{\Omega}$ is identical to the angular velocity $\bar{\omega}$ of the rigid body.

As in the case of planar kinematics, the vector $\bar{\omega}$ is independent of the choice of body-fixed coordinate system (*cf.* Theorem 12.5). However, this proof is omitted.

Definition 16.2 of angular velocity for general rigid-body motion is consistent with Def. 12.4 of angular velocity for planar motion. Adopting the notation in Fig. 12.3 for motion in the xy plane, we have

$$
\begin{aligned}
\bar{e}_x &= \cos\theta\,\bar{e}_X + \sin\theta\,\bar{e}_Y, & \dot{\bar{e}}_x &= \dot{\theta}(-\sin\theta\,\bar{e}_X + \cos\theta\,\bar{e}_Y) = \dot{\theta}\,\bar{e}_y, \\
\bar{e}_y &= -\sin\theta\,\bar{e}_X + \cos\theta\,\bar{e}_Y, & \dot{\bar{e}}_y &= \dot{\theta}(-\cos\theta\,\bar{e}_X - \sin\theta\,\bar{e}_Y) = -\dot{\theta}\,\bar{e}_x, \\
\bar{e}_z &= \bar{e}_Z, & \dot{\bar{e}}_z &= \bar{0}.
\end{aligned}
$$

Substitution of these expressions into Eq. (16.1) with $\bar{\omega} = \bar{\Omega}$ gives

$$
\bar{\omega} = (-\dot{\theta}\,\bar{e}_x \cdot \bar{e}_z)\bar{e}_x + (\bar{0} \cdot \bar{e}_x)\bar{e}_y + (\dot{\theta}\,\bar{e}_y \cdot \bar{e}_y)\bar{e}_z = \dot{\theta}\,\bar{e}_z = \dot{\theta}\,\bar{e}_Z,
$$

which is identical to Def. 12.4 for the angular velocity in the planar case. Evidently, the angular velocity $\bar{\omega}$ has a clear interpretation for planar motion as the rate of change of the polar angle.

Definition 16.3 (Angular acceleration). The *angular acceleration* of a rigid body in general, three-dimensional motion is

$$
\bar{\alpha} \equiv \dot{\bar{\omega}}, \tag{16.3}
$$

where $\bar{\omega}$ is the angular velocity of the body.

16.2 Coriolis equation

We shall use three *lemmata*, that exploit the constant length of the basis vectors and their orthogonality, to formulate the *Coriolis equation*, which is a differentiation rule for vectors represented in time-varying coordinate systems.

Lemma 16.4. Let $\bar{e}_\lambda$ be an arbitrary unit vector with time-dependent direction. Then, it holds that

$$
\dot{\bar{e}}_\lambda \cdot \bar{e}_\lambda = 0. \tag{16.4}
$$

Proof. Time differentiation of the identity $\bar{e}_\lambda \cdot \bar{e}_\lambda = 1$ using the Product rule, Eq. (A.25b), gives

$$\dot{\bar{e}}_\lambda \cdot \bar{e}_\lambda + \bar{e}_\lambda \cdot \dot{\bar{e}}_\lambda = 0 \quad \Leftrightarrow$$
$$\dot{\bar{e}}_\lambda \cdot \bar{e}_\lambda = 0. \qquad \square$$

From Lemma 16.4, the time derivative of a unit vector is perpendicular to the vector itself (Fig. 16.3).

Lemma 16.5. Let $\{\bar{e}_x, \bar{e}_y, \bar{e}_z\}$ be the basis of an orthogonal coordinate system with time-dependent axis directions. Then, it holds that

Figure 16.3: It holds that $\dot{\bar{e}}_\lambda \perp \bar{e}_\lambda$.

$$\dot{\bar{e}}_x \cdot \bar{e}_y = -\dot{\bar{e}}_y \cdot \bar{e}_x, \quad \dot{\bar{e}}_y \cdot \bar{e}_z = -\dot{\bar{e}}_z \cdot \bar{e}_y, \quad \dot{\bar{e}}_z \cdot \bar{e}_x = -\dot{\bar{e}}_x \cdot \bar{e}_z. \quad (16.5)$$

Proof. Time differentiation of the orthogonality condition $\bar{e}_x \cdot \bar{e}_y = 0$ using the Product rule, Eq. (A.25b), gives

$$\dot{\bar{e}}_x \cdot \bar{e}_y + \bar{e}_x \cdot \dot{\bar{e}}_y = 0 \quad \Leftrightarrow$$
$$\dot{\bar{e}}_x \cdot \bar{e}_y = -\dot{\bar{e}}_y \cdot \bar{e}_x.$$

Using the same method for $\bar{e}_y \cdot \bar{e}_z = 0$ and $\bar{e}_z \cdot \bar{e}_x = 0$ proves this *lemma.* $\qquad \square$

Lemma 16.6 (Time differentiation of a basis vector). Let $\{\bar{e}_x, \bar{e}_y, \bar{e}_z\}$ be the basis of a rotating coordinate system with the angular velocity $\bar{\Omega}$ relative to a fixed coordinate system. Then, it holds that (Fig. 16.4)

$$\dot{\bar{e}}_x = \bar{\Omega} \times \bar{e}_x, \quad \dot{\bar{e}}_y = \bar{\Omega} \times \bar{e}_y, \quad \dot{\bar{e}}_z = \bar{\Omega} \times \bar{e}_z. \quad (16.6)$$

Proof. According to Eq. (16.1), we have

$$\bar{\Omega} = (\dot{\bar{e}}_y \cdot \bar{e}_z)\bar{e}_x + (\dot{\bar{e}}_z \cdot \bar{e}_x)\bar{e}_y + (\dot{\bar{e}}_x \cdot \bar{e}_y)\bar{e}_z.$$

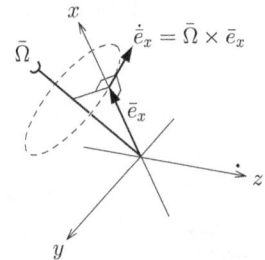

Figure 16.4: Geometry for Lemma 16.6 concerning the time differentiation of $\bar{e}_x$.

In the case of time differentiation of $\bar{e}_x$, the right-hand side in Eq. (16.6) becomes

$$\bar{\Omega} \times \bar{e}_x = (\dot{\bar{e}}_y \cdot \bar{e}_z)\bar{e}_x \times \bar{e}_x + (\dot{\bar{e}}_z \cdot \bar{e}_x)\bar{e}_y \times \bar{e}_x + (\dot{\bar{e}}_x \cdot \bar{e}_y)\bar{e}_z \times \bar{e}_x$$
$$= -(\dot{\bar{e}}_z \cdot \bar{e}_x)\bar{e}_z + (\dot{\bar{e}}_x \cdot \bar{e}_y)\bar{e}_y = \{\text{Lemma 16.5}\}$$
$$= (\dot{\bar{e}}_x \cdot \bar{e}_z)\bar{e}_z + (\dot{\bar{e}}_x \cdot \bar{e}_y)\bar{e}_y. \quad (16.7)$$

For the left-hand side, we can write $\dot{\bar{e}}_x$ as a sum of its components

$$\dot{\bar{e}}_x = (\dot{\bar{e}}_x \cdot \bar{e}_x)\bar{e}_x + (\dot{\bar{e}}_x \cdot \bar{e}_y)\bar{e}_y + (\dot{\bar{e}}_x \cdot \bar{e}_z)\bar{e}_z = \{\text{Lemma 16.4}\}$$
$$= (\dot{\bar{e}}_x \cdot \bar{e}_y)\bar{e}_y + (\dot{\bar{e}}_x \cdot \bar{e}_z)\bar{e}_z. \quad (16.8)$$

It follows from Eqs. (16.7) and (16.8) that $\dot{\bar{e}}_x = \bar{\Omega} \times \bar{e}_x$. Analogously, time differentiation of $\bar{e}_y$ and $\bar{e}_z$, respectively, proves this *lemma.* $\qquad \square$

Definition 16.7. Let $\bar{u} = u_x \bar{e}_x + u_y \bar{e}_y + u_z \bar{e}_z$ be an arbitrary time-dependent vector represented in a rotating coordinate system xyz. Then, the time derivative of $\bar{u}$ w.r.t. this rotating system is

$$\left.\frac{\mathrm{d}\bar{u}}{\mathrm{d}t}\right|_{xyz} \equiv \dot{u}_x \bar{e}_x + \dot{u}_y \bar{e}_y + \dot{u}_z \bar{e}_z. \tag{16.9}$$

However, the laws of motion must be applied in an inertial system, which is not permitted to rotate. One uses the Coriolis equation to differentiate a vector w.r.t. a fixed coordinate system:

Theorem 16.8 (The Coriolis equation). Let $\bar{u}$ be an arbitrary time-dependent vector, represented in a rotating coordinate system xyz with angular velocity $\bar{\Omega}$ relative to a fixed coordinate system. Then, it holds that

$$\dot{\bar{u}} = \left.\frac{\mathrm{d}\bar{u}}{\mathrm{d}t}\right|_{xyz} + \bar{\Omega} \times \bar{u}, \tag{16.10}$$

where $\dot{\bar{u}}$ is the time derivative of $\bar{u}$ relative to the fixed system, and $\mathrm{d}\bar{u}/\mathrm{d}t|_{xyz}$ is the time derivative of $\bar{u}$ relative to the rotating system.

Proof. The time-dependent vector $\bar{u}$ can be written

$$\bar{u} = u_x \bar{e}_x + u_y \bar{e}_y + u_z \bar{e}_z. \tag{16.11}$$

Time differentiation of Eq. (16.11) relative to the fixed coordinate system using the Product rule (A.25a) gives

$$\dot{\bar{u}} = \dot{u}_x \bar{e}_x + u_x \dot{\bar{e}}_x + \dot{u}_y \bar{e}_y + u_y \dot{\bar{e}}_y + \dot{u}_z \bar{e}_z + u_z \dot{\bar{e}}_z = \{\text{Eq. (16.9)}\}$$

$$= \left.\frac{\mathrm{d}\bar{u}}{\mathrm{d}t}\right|_{xyz} + u_x \dot{\bar{e}}_x + u_y \dot{\bar{e}}_y + u_z \dot{\bar{e}}_z = \{\text{Lemma 16.6}\}$$

$$= \left.\frac{\mathrm{d}\bar{u}}{\mathrm{d}t}\right|_{xyz} + u_x(\bar{\Omega} \times \bar{e}_x) + u_y(\bar{\Omega} \times \bar{e}_y) + u_z(\bar{\Omega} \times \bar{e}_z) = \{\text{Eq. (A.21b)}\}$$

$$= \left.\frac{\mathrm{d}\bar{u}}{\mathrm{d}t}\right|_{xyz} + \bar{\Omega} \times (u_x \bar{e}_x + u_y \bar{e}_y + u_z \bar{e}_z) = \{\text{Eq. (16.11)}\}$$

$$= \left.\frac{\mathrm{d}\bar{u}}{\mathrm{d}t}\right|_{xyz} + \bar{\Omega} \times \bar{u}. \qquad \square$$

16.3   Velocity and acceleration equations

For planar motion, we derived velocity and acceleration relations between two body-fixed points (Theorems 12.8 and 12.9). It will be shown that these relations still hold for general, three-dimensional motion.

Lemma 16.9 (Time differentiation of a body-fixed vector). For a rigid body with angular velocity $\bar{\omega}$, the time derivative of a body-fixed vector $\bar{u}$ is

$$\dot{\bar{u}} = \bar{\omega} \times \bar{u}. \tag{16.12}$$

Proof. Consider a body-fixed coordinate system xyz, having the angular velocity $\bar{\Omega} = \bar{\omega}$ relative to a fixed coordinate system (Fig. 16.5). Then, the Coriolis Eq. (16.10) gives

$$\dot{\bar{u}} = \left.\frac{d\bar{u}}{dt}\right|_{xyz} + \bar{\omega} \times \bar{u} = \{\bar{u} \text{ const. in } xyz\} = \bar{0} + \bar{\omega} \times \bar{u}. \qquad \square$$

Consequently, Lemma 12.7 for differentiation of body-fixed vectors, can be extended to general motion.

Theorem 16.10 (Relative-velocity equation). For a rigid body in general, three-dimensional motion with angular velocity $\bar{\omega}$, it holds that

$$\bar{v}_{\mathcal{Q}} = \bar{v}_{\mathcal{P}} + \bar{\omega} \times \overline{\mathcal{PQ}}, \qquad (16.13)$$

where $\mathcal{P}$ and $\mathcal{Q}$ are body-fixed points.

Proof. According to Lemma 16.9, we have $\dot{\overline{\mathcal{PQ}}} = \omega \times \overline{\mathcal{PQ}}$. We can reuse the proof of Theorem 12.8 to prove the present theorem. $\qquad \square$

Theorem 16.11 (Relative-acceleration equation). For a rigid body in general, three-dimensional motion with angular velocity $\bar{\omega}$ and angular acceleration $\bar{\alpha}$, it holds that

$$\bar{a}_{\mathcal{Q}} = \bar{a}_{\mathcal{P}} + \bar{\alpha} \times \overline{\mathcal{PQ}} + \bar{\omega} \times \left(\bar{\omega} \times \overline{\mathcal{PQ}}\right), \qquad (16.14)$$

where $\mathcal{P}$ and $\mathcal{Q}$ are body-fixed points.

Proof. According to Lemma 16.9, we have $\dot{\overline{\mathcal{PQ}}} = \omega \times \overline{\mathcal{PQ}}$. Then, it is possible to reuse the proof of Theorem 12.9 to prove this theorem. $\qquad \square$

16.4 Systems of rigid bodies

In a system of rigid bodies, enumerated $i = 0, 1, 2, \ldots, n$, where $i = 0$ represents a space-fixed foundation, every rigid body has its individual angular velocity $\bar{\omega}_i$ relative to a fixed coordinate system. Industrial robots are examples of such multi-body systems (Fig. 16.6). The engines or hydraulics that control the parts of the robot, affect the *relative* orientation of those parts. For this reason, it is usually the angular velocity $\bar{\omega}_{j/i}$ of body j relative to body i, that is directly accessible to the control system of the robot.

Definition 16.12 (Relative angular velocity). The *relative angular velocity* between a rigid body j and another rigid body i is

$$\bar{\omega}_{j/i} \equiv \bar{\omega}_j - \bar{\omega}_i, \qquad (16.15)$$

where $\bar{\omega}_i$ and $\bar{\omega}_j$ are the angular velocities of bodies i and j, respectively, relative to a fixed coordinate system.

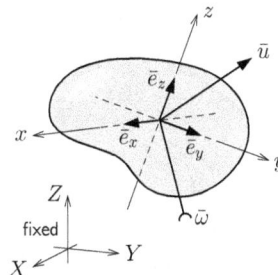

Figure 16.5: A body-fixed coordinate system xyz and a body-fixed vector $\bar{u}$ rotating with the angular velocity $\bar{\omega}$ relative to a space-fixed coordinate system XYZ.

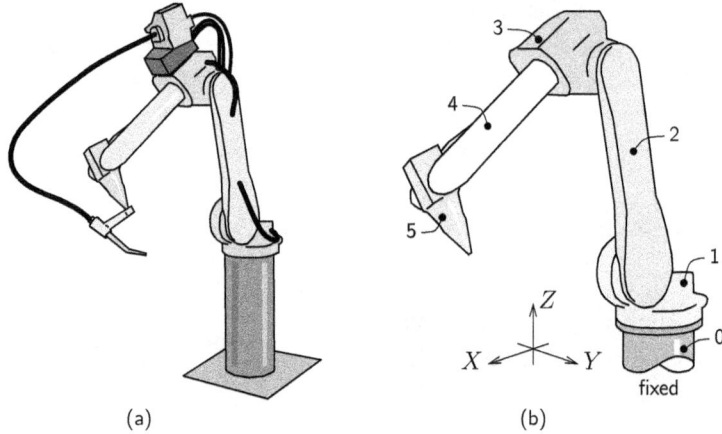

Figure 16.6: (a) An illustration of an arc welding robot. (b) The parts of the robot, $i = 0, 1, 2, 3, 4, 5$, are rigid bodies, rotating relative to each other.

Since $i = 0$ corresponds to the fixed system, we have $\bar{\omega}_0 = \bar{0}$, so that Def. 16.12 gives

$$\bar{\omega}_i = \bar{\omega}_{i/0}, \qquad i = 1, 2, \ldots, n. \tag{16.16}$$

Given the relative angular velocities of the rigid bodies in a mechanical system, we can calculate the angular velocities of these bodies relative to a fixed coordinate system.

Theorem 16.13. For a system of rigid bodies $i = 0, 1, 2, \ldots, n$, where $i = 0$ is a space-fixed body, it holds that

$$\bar{\omega}_n = \sum_{i=1}^{n} \bar{\omega}_{i/(i-1)} = \bar{\omega}_{n/(n-1)} + \cdots + \bar{\omega}_{2/1} + \bar{\omega}_{1/0}, \tag{16.17}$$

where $\bar{\omega}_n$ is the angular velocity of body n, and $\bar{\omega}_{i/(i-1)}$ is the relative angular velocity between bodies i and $i - 1$ (Fig. 16.7).

Proof. It follows from Def. 16.12 that

$$\sum_{i=1}^{n} \bar{\omega}_{i/(i-1)} = \sum_{i=1}^{n} (\bar{\omega}_i - \bar{\omega}_{i-1})$$

$$= \sum_{i=1}^{n} \bar{\omega}_i - \sum_{i=1}^{n} \bar{\omega}_{i-1} = \{\text{Index substitution}\}$$

$$= \sum_{i=1}^{n} \bar{\omega}_i - \sum_{i=0}^{n-1} \bar{\omega}_i$$

$$= \bar{\omega}_n + \sum_{i=1}^{n-1} \bar{\omega}_i - \sum_{i=1}^{n-1} \bar{\omega}_i - \bar{\omega}_0 = \{\bar{\omega}_0 = \bar{0}\}$$

$$= \bar{\omega}_n. \qquad \square$$

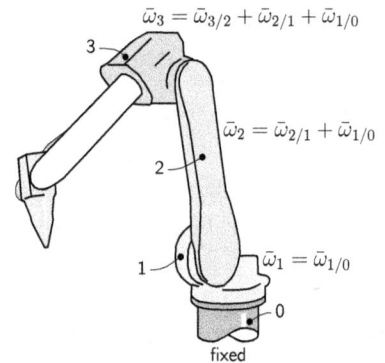

Figure 16.7: An angular velocity $\bar{\omega}_i$, relative to the fixed system, is the sum of relative angular velocities.

A relative angular velocity between two rigid bodies is sometimes called the *spin*, which we denote by $\bar{p}$. An illustrative example of spin, is the motion of a propeller relative to the body of an airplane. We introduce a coordinate system xyz, which is fixed relative to the fuselage (Body 1) with the x axis aligned with the propeller axle (Fig. 16.8). Then, Body 1 and its coordinate system have an angular velocity $\bar{\omega}_1 = \bar{\Omega}$ due to the maneuvering of the pilot. The engine speed p gives the propeller (Body 2) a spin $\bar{\omega}_{2/1} = \bar{p} = -p\bar{e}_x$, relative to the fuselage, so that the proper angular velocity of the propeller is

$$\bar{\omega}_2 = \bar{\omega}_{2/1} + \bar{\omega}_1 = \bar{p} + \bar{\Omega}.$$

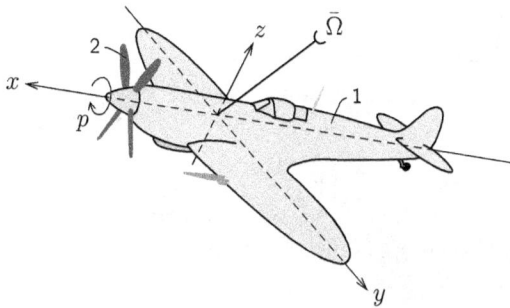

Figure 16.8: The propeller (2) has a spin $\bar{p} = -p\bar{e}_x$ relative to the body of the airplane (1), which in turn has the angular velocity $\bar{\omega}_1 = \bar{\Omega}$.

17

Three-dimensional kinetics of rigid bodies

Parts of Chapter 13 concerning planar kinetics treat rigid bodies of arbitrary shape in general, three-dimensional motion. This includes the Force equation,

$$\Sigma \bar{F} = \dot{\bar{G}} = m\bar{a}_{\mathcal{G}}, \tag{17.1}$$

and the Moment equations w.r.t. the center of mass $\mathcal{G}$ or a space-fixed point $\mathcal{D}$:

$$\Sigma \bar{M}_{\mathcal{G}} = \dot{\bar{H}}_{\mathcal{G}}, \tag{17.2a}$$

$$\Sigma \bar{M}_{\mathcal{D}} = \dot{\bar{H}}_{\mathcal{D}}. \tag{17.2b}$$

However, for general, three-dimensional rigid-body motion, the angular velocity can assume any arbitrary direction (Def. 16.2). For this reason, the expressions for angular momentum in planar motion do not hold in the three-dimensional case.

17.1   The inertia matrix

For planar motion, the moment of inertia was introduced (Def. 13.9) to describe the resistance of the rigid body against changing its angular velocity. For three-dimensional motion, the situation becomes more complicated due to the freedom of the axis of rotation to change its orientation. To address this, the inertia of the body is described by an *inertia matrix*. Its relation to motion, however, becomes clear in later Sect. 17.2. Herein, matrices are denoted by a double-bar over the variable name.

Definition 17.1 (Inertia matrix). The *inertia matrix*[27] of a rigid body Ω, w.r.t. an arbitrary point $\mathcal{A}$, is

$$\bar{\bar{I}}_{\mathcal{A}} \equiv \begin{bmatrix} I_{\mathcal{A}xx} & I_{\mathcal{A}xy} & I_{\mathcal{A}xz} \\ I_{\mathcal{A}xy} & I_{\mathcal{A}yy} & I_{\mathcal{A}yz} \\ I_{\mathcal{A}xz} & I_{\mathcal{A}yz} & I_{\mathcal{A}zz} \end{bmatrix}, \tag{17.3}$$

[27] Also called *inertia tensor*.

where $I_{\mathcal{A}xx}$, $I_{\mathcal{A}yy}$ and $I_{\mathcal{A}zz}$ are the *moments of inertia*, and $I_{\mathcal{A}xy}$, $I_{\mathcal{A}xz}$ and $I_{\mathcal{A}yz}$ are the *products of inertia*:

$$I_{\mathcal{A}xx} \equiv \int_\Omega (y^2 + z^2)\mathrm{d}m, \qquad I_{\mathcal{A}xy} \equiv -\int_\Omega xy\,\mathrm{d}m,$$

$$I_{\mathcal{A}yy} \equiv \int_\Omega (x^2 + z^2)\mathrm{d}m, \qquad I_{\mathcal{A}xz} \equiv -\int_\Omega xz\,\mathrm{d}m,$$

$$I_{\mathcal{A}zz} \equiv \int_\Omega (x^2 + y^2)\mathrm{d}m, \qquad I_{\mathcal{A}yz} \equiv -\int_\Omega yz\,\mathrm{d}m,$$

Figure 17.1: Geometry for Def. 17.1.

where $\bar{r} = x\bar{e}_x + y\bar{e}_y + z\bar{e}_z$ is a vector from $\mathcal{A}$ to the mass element $\mathrm{d}m$ (Fig. 17.1).[28]

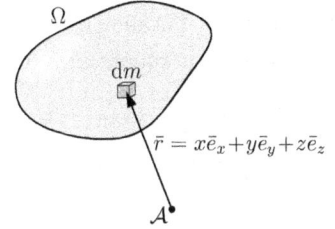

[28] In some presentations, the products of inertia are defined with reversed sign.

According to Def. 17.1, the inertia matrix is symmetric. The moments and products of inertia for a selection of simple bodies with uniformly distributed mass is found in Table C.2.

To calculate the inertia matrix w.r.t. a coordinate system xyz and a point $\mathcal{A}$, one must integrate the moments and products of inertia. The calculation of the products of inertia can be simplified for bodies with mirror symmetry w.r.t. one of the planes $x = 0$, $y = 0$ or $z = 0$. For instance, if the body is mirror-symmetric w.r.t. the plane $y = 0$, it follows that $I_{\mathcal{A}xy} = I_{\mathcal{A}yz} = 0$, since the integrands of these products of inertia are odd w.r.t. y (Fig. 17.2). For this particular example, we obtain

$$\bar{\bar{I}}_\mathcal{A} = \begin{bmatrix} I_{\mathcal{A}xx} & 0 & I_{\mathcal{A}xz} \\ 0 & I_{\mathcal{A}yy} & 0 \\ I_{\mathcal{A}xz} & 0 & I_{\mathcal{A}zz} \end{bmatrix}.$$

A similar line of reasoning applies for bodies that are mirror-symmetric w.r.t. the planes $x = 0$ or $z = 0$.

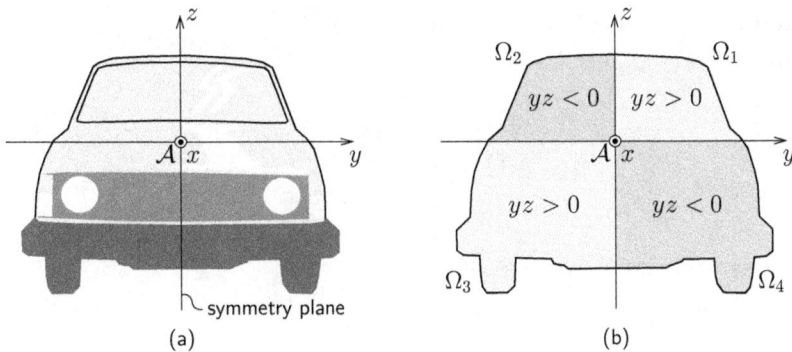

(a)

(b)

Figure 17.2: (a) With $\mathcal{A}$ as the origin, the vehicle is symmetric w.r.t. the plane $y = 0$. (b) Then, $\int_{\Omega_1} yz\,\mathrm{d}m = -\int_{\Omega_2} yz\,\mathrm{d}m$, and $\int_{\Omega_3} yz\,\mathrm{d}m = -\int_{\Omega_4} yz\,\mathrm{d}m$, which cancel out so that $I_{\mathcal{A}yz} = 0$. Similarly, $I_{\mathcal{A}xy} = 0$.

If the inertia matrix of a rigid body is known w.r.t. the center of mass, a transfer theorem can be used to calculate the inertia matrix w.r.t. any other point. We formulate this transfer theorem, first for moments of inertia and then for products of inertia.

Lemma 17.2 (Transfer theorem for moments of inertia). For a rigid body, the moments of inertia w.r.t. an arbitrary point $\mathcal{A}$ are

$$I_{\mathcal{A}xx} = I_{\mathcal{G}xx} + m(d_y^2 + d_z^2), \tag{17.4a}$$

$$I_{\mathcal{A}yy} = I_{\mathcal{G}yy} + m(d_x^2 + d_z^2), \tag{17.4b}$$

$$I_{\mathcal{A}zz} = I_{\mathcal{G}zz} + m(d_x^2 + d_y^2), \tag{17.4c}$$

where $\mathcal{G}$ is the center of mass of the body, m is its mass, and $\overline{\mathcal{A}\mathcal{G}} = \bar{d} = d_x \bar{e}_x + d_y \bar{e}_y + d_z \bar{e}_z$ is the *transfer vector*.

Proof. Place the origin in $\mathcal{A}$, and let $\bar{r} = x\bar{e}_x + y\bar{e}_y + z\bar{e}_z$ be the position vector of the mass element in the rigid body Ω. We write this position vector as $\bar{r} = \bar{d} + \bar{s}$, where $\bar{s}$ is a vector originating from $\mathcal{G}$ (Fig. 17.3). On component form, we obtain

$$x = d_x + s_x, \qquad y = d_y + s_y, \qquad z = d_z + s_z.$$

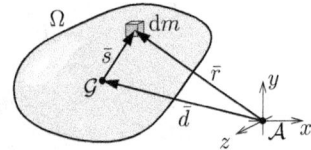

Figure 17.3: A rigid body where $\bar{s}$ is a vector from $\mathcal{G}$ to a mass element dm. The vector $\bar{r}$ is the position vector of the mass element relative to a coordinate system with its origin in $\mathcal{A}$.

According to Def. 17.1, we have

$$
\begin{aligned}
I_{\mathcal{A}zz} &= \int_\Omega (x^2 + y^2)\mathrm{d}m \\
&= \int_\Omega \left[(d_x + s_x)^2 + (d_y + s_y)^2 \right] \mathrm{d}m \\
&= \int_\Omega (d_x^2 + 2d_x s_x + s_x^2 + d_y^2 + 2d_y s_y + s_y^2)\mathrm{d}m \\
&= (d_x^2 + d_y^2)\underbrace{\int_\Omega \mathrm{d}m}_{=m} + 2d_x \underbrace{\int_\Omega s_x \mathrm{d}m}_{=0} + 2d_y \underbrace{\int_\Omega s_y \mathrm{d}m}_{=0} + \underbrace{\int_\Omega (s_x^2 + s_y^2)\mathrm{d}m}_{=I_{\mathcal{G}zz}} \\
&= I_{\mathcal{G}zz} + m(d_x^2 + d_y^2).
\end{aligned}
$$

Analogous reasoning for $I_{\mathcal{A}xx}$ and $I_{\mathcal{A}yy}$ proves this lemma. $\qquad\square$

Lemma 17.3 (Transfer theorem for products of inertia). For a rigid body, the products of inertia w.r.t. an arbitrary point $\mathcal{A}$ are

$$I_{\mathcal{A}xy} = I_{\mathcal{G}xy} - md_x d_y, \tag{17.5a}$$

$$I_{\mathcal{A}xz} = I_{\mathcal{G}xz} - md_x d_z, \tag{17.5b}$$

$$I_{\mathcal{A}yz} = I_{\mathcal{G}yz} - md_y d_z, \tag{17.5c}$$

where $\mathcal{G}$ is the center of mass of the body, m is its mass, and $\overline{\mathcal{A}\mathcal{G}} = \bar{d} = d_x \bar{e}_x + d_y \bar{e}_y + d_z \bar{e}_z$ is the transfer vector.

Proof. Place the origin in $\mathcal{A}$, and let $\bar{r} = x\bar{e}_x + y\bar{e}_y + z\bar{e}_z$ be the position vector of the mass element in the rigid body Ω. We have $\bar{r} = \bar{d} + \bar{s}$, where $\bar{s}$ is a vector originating from $\mathcal{G}$ (Fig. 17.3). On component form, we have

$$x = d_x + s_x, \qquad y = d_y + s_y, \qquad z = d_z + s_z.$$

According to Def. 17.1, we have

$$
\begin{aligned}
I_{\mathcal{A}xy} &= -\int_{\Omega} xy\,\mathrm{d}m \\
&= -\int_{\Omega} (d_x + s_x)(d_y + s_y)\,\mathrm{d}m \\
&= -\int_{\Omega} d_x d_y + d_x s_y + d_y s_x + s_x s_y\,\mathrm{d}m \\
&= -d_x d_y \underbrace{\int_{\Omega} \mathrm{d}m}_{=m} - d_x \underbrace{\int_{\Omega} s_y\,\mathrm{d}m}_{=0} - d_y \underbrace{\int_{\Omega} s_x\,\mathrm{d}m}_{=0} - \underbrace{\int_{\Omega} s_x s_y\,\mathrm{d}m}_{=I_{\mathcal{G}xy}} \\
&= I_{\mathcal{G}xy} - m d_x d_y.
\end{aligned}
$$

Analogous reasoning for $I_{\mathcal{A}xz}$ and $I_{\mathcal{A}yz}$ proves this lemma. $\square$

Theorem 17.4 (Parallel axis theorem). The inertia matrix of a rigid body w.r.t. an arbitrary point $\mathcal{A}$ is

$$
\bar{\bar{I}}_{\mathcal{A}} = \bar{\bar{I}}_{\mathcal{G}} + m
\begin{bmatrix}
d_y^2 + d_z^2 & -d_x d_y & -d_x d_z \\
-d_x d_y & d_x^2 + d_z^2 & -d_y d_z \\
-d_x d_z & -d_y d_z & d_x^2 + d_y^2
\end{bmatrix},
\tag{17.6}
$$

where $\mathcal{G}$ is the center of mass of the body, m is its mass, and $\overline{\mathcal{AG}} = \bar{d} = d_x \bar{e}_x + d_y \bar{e}_y + d_z \bar{e}_z$ is the transfer vector (Fig. 17.3).

Proof. The validity of Eq. (17.6) is clear from Lemma 17.2 for the diagonal elements, and from Lemma 17.3 for the off-diagonal elements. $\square$

In problem solving, one can frequently find a tabulated value for the inertia matrix $\bar{\bar{I}}_{\mathcal{G}}$ w.r.t. the center of mass $\mathcal{G}$ (Table C.2), and then use Eq. (17.6) to calculate the inertia matrix w.r.t. a desired point $\mathcal{A}$.

17.2   Angular momentum

The expression for angular momentum in Def. 13.2 can be simplified for rigid bodies in general, three-dimensional motion. We prepare for this by formulating two *lemmata*.

Lemma 17.5. The angular momentum w.r.t. the center of mass $\mathcal{G}$ of a rigid body in general, three-dimensional motion is

$$
\bar{H}_{\mathcal{G}} = \int_{\Omega} \bar{s} \times (\bar{\omega} \times \bar{s})\,\mathrm{d}m,
\tag{17.7}
$$

where $\bar{s}$ is a vector from $\mathcal{G}$ to the mass element $\mathrm{d}m$, and $\bar{\omega}$ is the angular velocity of the body.

Proof. We follow the proof in Lemma 13.11, only with the difference that the relation $\dot{\bar{s}} = \bar{\omega} \times \bar{s}$ relies on Lemma 16.9 for general motion. $\square$

Lemma 17.6. For two arbitrary vectors $\bar{u}$ and $\bar{w}$, it holds that

$$\bar{u} \times (\bar{w} \times \bar{u}) = \begin{bmatrix} u_y^2 + u_z^2 & -u_x u_y & -u_x u_z \\ -u_x u_y & u_x^2 + u_z^2 & -u_y u_z \\ -u_x u_z & -u_y u_z & u_x^2 + u_y^2 \end{bmatrix} \bar{w}. \qquad (17.8)$$

Proof. According to Eq. (A.22a), we have

$$\bar{u} \times (\bar{w} \times \bar{u}) = (\bar{u} \cdot \bar{u})\bar{w} - (\bar{u} \cdot \bar{w})\bar{u}$$

$$= \begin{bmatrix} (u_x^2 + u_y^2 + u_z^2)w_x - (u_x w_x + u_y w_y + u_z w_z)u_x \\ (u_x^2 + u_y^2 + u_z^2)w_y - (u_x w_x + u_y w_y + u_z w_z)u_y \\ (u_x^2 + u_y^2 + u_z^2)w_z - (u_x w_x + u_y w_y + u_z w_z)u_z \end{bmatrix}$$

$$= \begin{bmatrix} (u_y^2 + u_z^2)w_x - u_x u_y w_y - u_x u_z w_z \\ -u_x u_y w_x + (u_x^2 + u_z^2)w_y - u_y u_z w_z \\ -u_x u_z w_x - u_y u_z w_y + (u_x^2 + u_y^2)w_z \end{bmatrix}$$

$$= \begin{bmatrix} u_y^2 + u_z^2 & -u_x u_y & -u_x u_z \\ -u_x u_y & u_x^2 + u_z^2 & -u_y u_z \\ -u_x u_z & -u_y u_z & u_x^2 + u_y^2 \end{bmatrix} \bar{w}. \qquad \square$$

Using Lemmata 17.5 and 17.6, we can derive an expression for the angular momentum of a rigid body, w.r.t. its center of mass.

Theorem 17.7. For a rigid body in general, three-dimensional motion, the angular momentum w.r.t. the center of mass $\mathcal{G}$ is

$$\bar{H}_\mathcal{G} = \bar{\bar{I}}_\mathcal{G}\bar{\omega}, \qquad (17.9)$$

where $\bar{\omega}$ is the angular velocity of the rigid body, and $\bar{\bar{I}}_\mathcal{G}$ is the inertia matrix of this body w.r.t. $\mathcal{G}$.

Proof. Let $\bar{s} = s_x \bar{e}_x + s_y \bar{e}_y + s_z \bar{e}_z$ be a vector from $\mathcal{G}$ to a mass element $\mathrm{d}m$ in the rigid body Ω. Then, Eq. (17.7) gives

$$\bar{H}_\mathcal{G} = \int_\Omega \bar{s} \times (\bar{\omega} \times \bar{s})\mathrm{d}m = \{\text{Eq. } (17.8)\}$$

$$= \int_\Omega \begin{bmatrix} s_y^2 + s_z^2 & -s_x s_y & -s_x s_z \\ -s_x s_y & s_x^2 + s_z^2 & -s_y s_z \\ -s_x s_z & -s_y s_z & s_x^2 + s_y^2 \end{bmatrix} \bar{\omega}\mathrm{d}m = \{\bar{\omega} \text{ const.}\}$$

$$= \begin{bmatrix} \int_\Omega (s_y^2 + s_z^2)\mathrm{d}m & -\int_\Omega s_x s_y \mathrm{d}m & -\int_\Omega s_x s_z \mathrm{d}m \\ -\int_\Omega s_x s_y \mathrm{d}m & \int_\Omega (s_x^2 + s_z^2)\mathrm{d}m & -\int_\Omega s_y s_z \mathrm{d}m \\ -\int_\Omega s_x s_z \mathrm{d}m & -\int_\Omega s_y s_z \mathrm{d}m & \int_\Omega (s_x^2 + s_y^2)\mathrm{d}m \end{bmatrix} \bar{\omega} = \{\text{Def. } 17.1\}$$

$$= \begin{bmatrix} I_{\mathcal{G}xx} & I_{\mathcal{G}xy} & I_{\mathcal{G}xz} \\ I_{\mathcal{G}xy} & I_{\mathcal{G}yy} & I_{\mathcal{G}yz} \\ I_{\mathcal{G}xz} & I_{\mathcal{G}yz} & I_{\mathcal{G}zz} \end{bmatrix} \bar{\omega} = \{\text{Def. } 17.1\}$$

$$= \bar{\bar{I}}_\mathcal{G}\bar{\omega}. \qquad \square$$

Since the angular momentum of the rigid body is $\bar{H}_{\mathcal{G}} = \bar{\bar{I}}_{\mathcal{G}}\bar{\omega}$, we understand that $\bar{H}_{\mathcal{G}}$ is not necessarily parallel to $\bar{\omega}$ (Fig. 17.4). This is an important difference as compared to the situation for flat, rigid bodies in planar motion.

Theorem 17.8. For a rigid body in general, three-dimensional motion, the angular momentum w.r.t. a body- and space-fixed point $\mathcal{O}$ is

$$\bar{H}_{\mathcal{O}} = \bar{\bar{I}}_{\mathcal{O}}\bar{\omega}, \tag{17.10}$$

where $\bar{\omega}$ is the angular velocity of the body, and $\bar{\bar{I}}_{\mathcal{O}}$ is the inertia matrix of the body w.r.t. $\mathcal{O}$.

Proof. We choose $\mathcal{O}$ as the origin, so that $\bar{d} = \overline{\mathcal{OG}}$ is both a position vector for $\mathcal{G}$, and a body-fixed vector. Then, Lemma 16.9 gives

$$\bar{v}_{\mathcal{G}} = \dot{\bar{d}} = \bar{\omega} \times \bar{d}.$$

According to the Transfer theorem of angular momentum 13.6, we have

$$\begin{aligned}
\bar{H}_{\mathcal{O}} &= \bar{H}_{\mathcal{G}} + \bar{d} \times \bar{G} = \left\{ \text{Eqs. (13.5) and (17.9)} \right\} \\
&= \bar{\bar{I}}_{\mathcal{G}}\bar{\omega} + \bar{d} \times m\bar{v}_{\mathcal{G}} \\
&= \bar{\bar{I}}_{\mathcal{G}}\bar{\omega} + m\bar{d} \times (\bar{\omega} \times \bar{d}) = \left\{ \text{Eq. (17.8)} \right\} \\
&= \bar{\bar{I}}_{\mathcal{G}}\bar{\omega} + m \begin{bmatrix} d_y^2 + d_z^2 & -d_x d_y & -d_x d_z \\ -d_x d_y & d_x^2 + d_z^2 & -d_y d_z \\ -d_x d_z & -d_y d_z & d_x^2 + d_y^2 \end{bmatrix} \bar{\omega} = \left\{ \text{Eq. (17.6)} \right\} \\
&= \bar{\bar{I}}_{\mathcal{O}}\bar{\omega}. \qquad \square
\end{aligned}$$

These expressions derived for $\bar{H}_{\mathcal{G}}$ and $\bar{H}_{\mathcal{O}}$ can be used in the respective Moment equations, Eqs. (17.2a) and (17.2b). This gives a relation between the moment sum of the force system and the rotational motion of the rigid body.

The momentum and the angular momentum of a rigid body appear in the expression for kinetic energy of a rigid body in general, three-dimensional motion.

Theorem 17.9 (Kinetic energy). The kinetic energy of a rigid body in general, three-dimensional motion is

$$K = \frac{1}{2}\bar{v}_{\mathcal{G}} \cdot \bar{G} + \frac{1}{2}\bar{\omega} \cdot \bar{H}_{\mathcal{G}}, \tag{17.11}$$

where $\bar{v}_{\mathcal{G}}$ is the velocity of the center of mass $\mathcal{G}$, $\bar{G}$ is the momentum of the body, $\bar{H}_{\mathcal{G}}$ is its angular momentum w.r.t. $\mathcal{G}$, and $\bar{\omega}$ is its angular velocity.

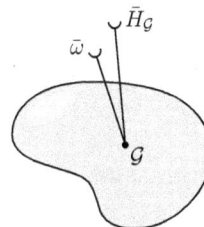

Figure 17.4: The direction of the angular momentum $\bar{H}_{\mathcal{G}} = \bar{\bar{I}}_{\mathcal{G}}\bar{\omega}$ is not necessarily the same as that of the angular velocity $\bar{\omega}$.

Proof. Let $\bar{r}_{\mathcal{G}}$ be the position vector of $\mathcal{G}$, so that the position vector $\bar{r}$ for a mass element in the body Ω is (Fig. 17.3)

$$\bar{r} = \bar{r}_{\mathcal{G}} + \bar{s} \quad \Rightarrow \quad \bar{v} = \bar{v}_{\mathcal{G}} + \dot{\bar{s}} \quad \Leftrightarrow \quad \bar{v} = \bar{v}_{\mathcal{G}} + \bar{\omega} \times \bar{s},$$

where Theorem 16.9 gave that $\dot{\bar{s}} = \bar{\omega} \times \bar{s}$. According to Def. 14.7, we obtain

$$
\begin{aligned}
K &= \frac{1}{2} \int_\Omega (\bar{v} \cdot \bar{v}) \mathrm{d}m \\
&= \frac{1}{2} \int_\Omega (\bar{v}_{\mathcal{G}} + \bar{\omega} \times \bar{s}) \cdot (\bar{v}_{\mathcal{G}} + \bar{\omega} \times \bar{s}) \, \mathrm{d}m \\
&= \frac{1}{2} \int_\Omega [\bar{v}_{\mathcal{G}} \cdot \bar{v}_{\mathcal{G}} + 2\bar{v}_{\mathcal{G}} \cdot (\bar{\omega} \times \bar{s}) + (\bar{\omega} \times \bar{s}) \cdot (\bar{\omega} \times \bar{s})] \, \mathrm{d}m = \{\text{Eq. (A.22b)}\} \\
&= \frac{1}{2} \int_\Omega \{\bar{v}_{\mathcal{G}} \cdot \bar{v}_{\mathcal{G}} + 2\bar{v}_{\mathcal{G}} \cdot (\bar{\omega} \times \bar{s}) + \bar{\omega} \cdot [\bar{s} \times (\bar{\omega} \times \bar{s})]\} \, \mathrm{d}m = \{\bar{\omega}, \bar{v}_{\mathcal{G}} \text{ const.}\} \\
&= \frac{1}{2} (\bar{v}_{\mathcal{G}} \cdot \bar{v}_{\mathcal{G}}) \underbrace{\int_\Omega \mathrm{d}m}_{=m} + \bar{v}_{\mathcal{G}} \cdot \left(\bar{\omega} \times \underbrace{\int_\Omega \bar{s} \mathrm{d}m}_{=\bar{0}} \right) + \frac{1}{2} \bar{\omega} \cdot \int_\Omega \bar{s} \times (\bar{\omega} \times \bar{s}) \mathrm{d}m = \{\text{Lemma 17.5}\} \\
&= \frac{1}{2} \bar{v}_{\mathcal{G}} \cdot \bar{G} + \frac{1}{2} \bar{\omega} \cdot \bar{H}_{\mathcal{G}}. \hspace{3cm} \square
\end{aligned}
$$

The Work–energy theorem, Eq. (14.17), still holds for general, three-dimensional motion, where the kinetic energy is calculated using Theorem 17.9.

17.3   Dynamic phenomena

It is possible to combine the laws of kinematics, the Force equation, the Moment equation, and an expression for angular momentum to solve dynamics problems that include general, three-dimensional motion. Two phenomena of particular importance are *dynamic unbalance* and *gyrodynamics*. Herein, we study examples to demonstrate how these problem classes can be addressed.

Dynamic unbalance

A rigid body that rotates about a fixed axis may exert forces and couples on its supports due to *dynamic unbalance*.

Consider the mechanical system in Fig. 17.5. A massless axle is mounted between bearings at $\mathcal{O}$ and $\mathcal{A}$. A slender cylinder, with length ℓ and mass m, is rigidly attached to the axle $\mathcal{OA}$ by means of a massless rod of length b. A couple C acts on $\mathcal{OA}$ and drives the rotation of this composite rigid body. The bearings allow for all forms of rotation, but constrain displacement in the radial directions.[29] The bearing at $\mathcal{O}$ also constrains axial displacement. We seek the reaction force $\bar{F}_{\mathcal{A}}$ at $\mathcal{A}$ for a given instantaneous angular velocity ω. The force of gravity is neglected.

[29] *Radial direction* – any direction perpendicular to the axial direction of an object or system.

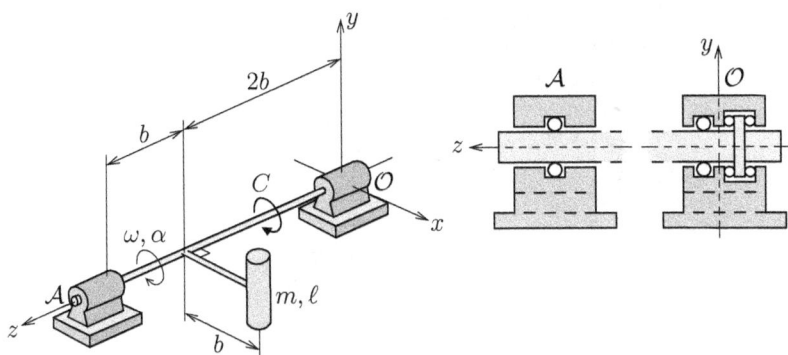

Figure 17.5: A slender cylinder of mass m is attached to a rigid, massless axle $\mathcal{OA}$ supported by two bearings. Rotations are permitted by the bearings, but displacement in the radial direction is constrained. The bearing $\mathcal{O}$ also constrains axial displacement.

Choice of coordinate system: We introduce a coordinate system xyz, so that the z axis coincides with $\mathcal{OA}$, and so that this coordinate system rotates with the axle $\mathcal{OA}$. The angular velocity of this axle is $\bar{\omega} = -\omega\bar{e}_z$, and its angular acceleration is $\bar{\alpha} = -\alpha\bar{e}_z$, where α is unknown. Thus, the angular velocityÂ of the coordinate system is $\bar{\Omega} = \bar{\omega}$.

Free-body diagram: We draw a free-body diagram for the body in an arbitrary position, while considering the rotating coordinate system:

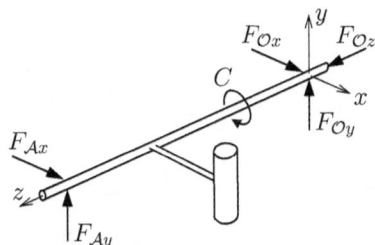

Equations of motion: We employ the Moment equation (17.2b) w.r.t. the body- and space-fixed point $\mathcal{O}$, so that the unknown force components at $\mathcal{O}$ are eliminated in the moment sum:

$$\Sigma\bar{M}_{\mathcal{O}} = \dot{\bar{H}}_{\mathcal{O}}. \tag{17.12}$$

Moment sum: The moment sum w.r.t. $\mathcal{O}$, Eq. (2.10), is:

$$\begin{aligned}
\Sigma\bar{M}_{\mathcal{O}} &= \overline{\mathcal{OA}} \times \bar{F}_{\mathcal{A}} + \bar{C} \\
&= 3b\bar{e}_z \times (F_{\mathcal{A}x}\bar{e}_x + F_{\mathcal{A}y}\bar{e}_y) - C\bar{e}_z \\
&= 3bF_{\mathcal{A}x}\bar{e}_y - 3bF_{\mathcal{A}y}\bar{e}_x - C\bar{e}_z.
\end{aligned}$$

Angular momentum: To calculate $\dot{\bar{H}}_\mathcal{O}$ on the right-hand side of Eq. (17.12), we first need to determine $\bar{H}_\mathcal{O} = \bar{\bar{I}}_\mathcal{O}\bar{\omega}$. The transfer vector of the massive cylinder is:

$$\bar{d} = \overline{\mathcal{O}\mathcal{G}} = \begin{bmatrix} b \\ 0 \\ 2b \end{bmatrix}.$$

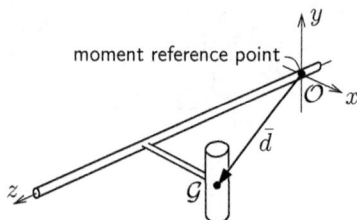

moment reference point

We approximate the cylinder by a slender bar of length ℓ. Its inertia matrix is given by the Parallel axis theorem 17.4:

$$\bar{\bar{I}}_\mathcal{O} = \bar{\bar{I}}_\mathcal{G} + m \begin{bmatrix} d_y^2 + d_z^2 & -d_x d_y & -d_x d_z \\ -d_x d_y & d_x^2 + d_z^2 & -d_y d_z \\ -d_x d_z & -d_y d_z & d_x^2 + d_y^2 \end{bmatrix} = \left\{ \begin{array}{c} \text{Table C.2,} \\ \text{cylinder, } r \to 0 \end{array} \right\}$$

$$= \begin{bmatrix} \frac{1}{12}m\ell^2 & 0 & 0 \\ 0 & 0 & 0 \\ 0 & 0 & \frac{1}{12}m\ell^2 \end{bmatrix} + m \begin{bmatrix} 4b^2 & 0 & -2b^2 \\ 0 & 5b^2 & 0 \\ -2b^2 & 0 & b^2 \end{bmatrix}.$$

Then, we calculate the angular momentum:

$$\bar{H}_\mathcal{O} = \bar{\bar{I}}_\mathcal{O}\bar{\omega}$$

$$= \begin{bmatrix} \frac{1}{12}m\ell^2 + 4mb^2 & 0 & -2mb^2 \\ 0 & 5mb^2 & 0 \\ -2mb^2 & 0 & \frac{1}{12}m\ell^2 + mb^2 \end{bmatrix} \begin{bmatrix} 0 \\ 0 \\ -\omega \end{bmatrix}$$

$$= \underbrace{2mb^2\omega}_{=H_{\mathcal{O}x}}\,\bar{e}_x + \underbrace{\left(-\frac{1}{12}m\ell^2 - mb^2\right)\omega}_{=H_{\mathcal{O}z}}\,\bar{e}_z.$$

Time derivative of angular momentum: Since the expression for the angular momentum contains time-varying basis vectors, the Coriolis equation (16.10) is used for differentiation:

$$\dot{\bar{H}}_\mathcal{O} = \left.\frac{\mathrm{d}\bar{H}_\mathcal{O}}{\mathrm{d}t}\right|_{xyz} + \bar{\Omega} \times \bar{H}_\mathcal{O} = \{\bar{\Omega} = \bar{\omega}\}$$

$$= \dot{H}_{\mathcal{O}x}\bar{e}_x + \dot{H}_{\mathcal{O}y}\bar{e}_y + \dot{H}_{\mathcal{O}z}\bar{e}_z + \bar{\omega} \times \bar{H}_\mathcal{O}$$

$$= 2mb^2\dot{\omega}\bar{e}_x - \left(\frac{1}{12}m\ell^2 + mb^2\right)\dot{\omega}\bar{e}_z$$

$$+ (-\omega\bar{e}_z) \times \left[2mb^2\omega\bar{e}_x - \left(\frac{1}{12}m\ell^2 + mb^2\right)m\omega\bar{e}_z\right]$$

$$= 2mb^2\alpha\bar{e}_x - \left(\frac{1}{12}m\ell^2 + mb^2\right)\alpha\bar{e}_z - 2mb^2\omega^2\bar{e}_y.$$

Calculations: Substitution of the expressions for $\Sigma \bar{M}_{\mathcal{O}}$ and $\dot{\bar{H}}_{\mathcal{O}}$ into Eq. (17.12), and identification of its x, y and z components, yield a system of equations:

$$-3bF_{Ay} = 2mb^2\alpha,$$
$$3bF_{Ax} = -2mb^2\omega^2,$$
$$-C = -\left(\frac{1}{12}m\ell^2 + mb^2\right)\alpha.$$

This system of equations is solved for F_{Ax} and F_{Ay}, which gives the solution:

$$\bar{F}_A = -\frac{2}{3}mb\omega^2\bar{e}_x - \frac{8bC}{12b^2 + \ell^2}\bar{e}_y. \qquad \square$$

Note that, since the directions of the basis vectors $\bar{e}_x$ and $\bar{e}_y$ vary with time, the direction of the reaction force $\bar{F}_A$ varies with time. Further analysis using the Force equation (17.1) would yield an expression for the reaction force $\bar{F}_{\mathcal{O}}$ from bearing $\mathcal{O}$.

Gyrodynamics

A gyroscope is a spinning body, such that its spin $\bar{p}$ is free to change direction in space. It is difficult to get an intuitive sense of the mechanical properties of a gyroscope. Therefore, it is necessary to rely on the governing equations, as seen in the example below.

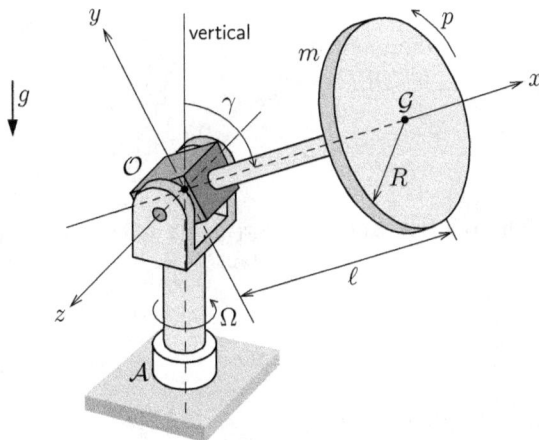

Figure 17.6: A circular disc with mass m and spin p is attached *via* a massless axle $\mathcal{OG}$ to a mechanism, which allows for free rotation of $\mathcal{OG}$ about a vertical line as well as the z axis.

Consider the mechanical system in Fig. 17.6. A massless axle $\mathcal{AO}$ is free to rotate around a vertical line owing to a bearing at $\mathcal{A}$. Another massless axle $\mathcal{OG}$ is attached to $\mathcal{AO}$ with a fork joint at $\mathcal{O}$, and $\mathcal{OG}$ forms an angle γ with the vertical line. A homogeneous circular disc with mass m, radius R, and center of mass $\mathcal{G}$ is welded onto $\mathcal{OG}$. The joint at $\mathcal{O}$ have bearings such that the joint does not exert any couple on $\mathcal{OG}$ in the axial direction. This allows for the disc to rotate with a spin p relative to the connecting body at $\mathcal{O}$. Even though this mechanical system allows for the angle γ to vary, we tentatively assume that $\dot{\gamma} = 0$ when analyzing the motion of the circular disc.

Choice of coordinate system: We introduce a coordinate system xyz, such that the x direction coincides with $\mathcal{OG}$, and the z direction is at all times horizontal. Thus, this coordinate system rotates with the intermediate body that connects the fork joint with the axle of the disc (Fig. 17.6).

Free-body diagram: We draw free-body diagrams for the axle $\mathcal{AO}$ and for the disc with its axle in an arbitrary position, while considering the rotating coordinate system. The bearings at $\mathcal{O}$ ensure that $\bar{C}_\mathcal{O} = C_{\mathcal{O}y}\bar{e}_y$.

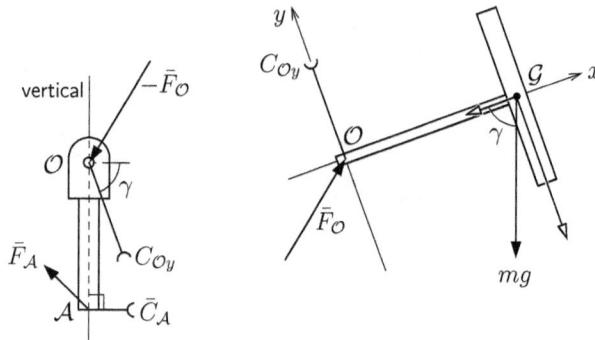

Kinematic relations: The three bodies of the system are enumerated according to Fig. 17.7. The axle $\mathcal{AO}$ has an angular velocity directed along the vertical line:

$$\bar{\omega}_1 = \Omega(\cos\gamma\,\bar{e}_x + \sin\gamma\,\bar{e}_y),$$

where $\Omega = \Omega(t)$ is an unknown function of time. The relative angular velocities of the system are given by

$$\bar{\omega}_{2/1} = -\dot{\gamma}\bar{e}_z = \{\text{assumption } \dot{\gamma} = 0\} = \bar{0},$$
$$\bar{\omega}_{3/2} = p\bar{e}_x.$$

Since the coordinate system xyz is fixed to Body 2, its angular velocity is

$$\bar{\Omega} = \bar{\omega}_2 = \{\text{Theorem 16.13}\} = \bar{\omega}_{2/1} + \bar{\omega}_1 = \Omega(\cos\gamma\,\bar{e}_x + \sin\gamma\,\bar{e}_y),$$

while the angular velocity of the disc is

$$\bar{\omega}_3 = \{\text{Theorem 16.13}\} = \bar{\omega}_{3/2} + \bar{\omega}_{2/1} + \bar{\omega}_1 = (p + \Omega\cos\gamma)\bar{e}_x + \Omega\sin\gamma\,\bar{e}_y.$$

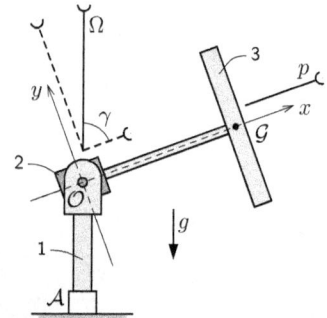

Figure 17.7: The system in Fig. 17.6 contains three bodies $i = 1, 2, 3$.

Equations of motion: A moment equilibrium for the massless axle, Body 1, w.r.t. the vertical line in the free-body diagram gives $C_{\mathcal{O}y} = 0$. For the disc, Body 3, we employ the Moment equation w.r.t. the body- and space-fixed point $\mathcal{O}$, Eq. (17.2b), so that the unknown force vector at $\mathcal{O}$ is eliminated:

$$\Sigma \bar{M}_{\mathcal{O}} = \dot{\bar{H}}_{\mathcal{O}}. \tag{17.13}$$

Moment sum: The moment sum w.r.t. $\mathcal{O}$, Eq. (2.10), is

$$
\begin{aligned}
\Sigma \bar{M}_{\mathcal{O}} &= \overline{\mathcal{OG}} \times m\bar{g} + \bar{C}_{\mathcal{O}} \\
&= \ell \bar{e}_x \times mg(-\cos\gamma\, \bar{e}_x - \sin\gamma\, \bar{e}_y) + C_{\mathcal{O}y}\bar{e}_y \\
&= -mg\ell \sin\gamma\, \bar{e}_z.
\end{aligned}
$$

Angular momentum: To calculate $\dot{\bar{H}}_{\mathcal{O}}$ on the right-hand side of the Moment equation (17.13), we first need to determine $\bar{H}_{\mathcal{O}} = \bar{\bar{I}}_{\mathcal{O}}\bar{\omega}_3$. The transfer vector of the disc is

$$\bar{d} = \overline{\mathcal{OG}} = \begin{bmatrix} \ell \\ 0 \\ 0 \end{bmatrix}.$$

moment reference point

According to the Parallel axis theorem 17.4, the inertia matrix is

$$
\bar{\bar{I}}_{\mathcal{O}} = \bar{\bar{I}}_{\mathcal{G}} + m \begin{bmatrix} d_y^2 + d_z^2 & -d_x d_y & -d_x d_z \\ -d_x d_y & d_x^2 + d_z^2 & -d_y d_z \\ -d_x d_z & -d_y d_z & d_x^2 + d_y^2 \end{bmatrix} = \left\{ \begin{array}{l} \text{Table C.2,} \\ \text{cylinder, } \ell \to 0 \end{array} \right\}
$$

$$
= \begin{bmatrix} \frac{1}{2}mR^2 & 0 & 0 \\ 0 & \frac{1}{4}mR^2 & 0 \\ 0 & 0 & \frac{1}{4}mR^2 \end{bmatrix} + m \begin{bmatrix} 0 & 0 & 0 \\ 0 & \ell^2 & 0 \\ 0 & 0 & \ell^2 \end{bmatrix}.
$$

This gives the angular momentum

$$
\begin{aligned}
\bar{H}_{\mathcal{O}} &= \bar{\bar{I}}_{\mathcal{O}}\bar{\omega}_3 \\
&= \begin{bmatrix} \frac{1}{2}mR^2 & 0 & 0 \\ 0 & \frac{1}{4}mR^2 + m\ell^2 & 0 \\ 0 & 0 & \frac{1}{4}mR^2 + m\ell^2 \end{bmatrix} \begin{bmatrix} p + \Omega\cos\gamma \\ \Omega\sin\gamma \\ 0 \end{bmatrix} \\
&= \underbrace{\frac{1}{2}mR^2 \left(p + \Omega\cos\gamma \right) \bar{e}_x}_{=H_{\mathcal{O}x}} + \underbrace{\left(\frac{1}{4}mR^2 + m\ell^2 \right) \Omega\sin\gamma\, \bar{e}_y}_{=H_{\mathcal{O}y}}.
\end{aligned}
$$

Time derivative of angular momentum: Since the expression for the angular momentum contains time-varying basis vectors, we use the Coriolis equation (16.10) for time differentiation:

$$\dot{\bar{H}}_{\mathcal{O}} = \left.\frac{\mathrm{d}\bar{H}_{\mathcal{O}}}{\mathrm{d}t}\right|_{xyz} + \bar{\Omega} \times \bar{H}_{\mathcal{O}}$$

$$= \dot{H}_{\mathcal{O}x}\bar{e}_x + \dot{H}_{\mathcal{O}y}\bar{e}_y + \dot{H}_{\mathcal{O}z}\bar{e}_z + \bar{\Omega} \times \bar{H}_{\mathcal{O}} = \{\cdots\}$$

$$= \frac{1}{2}mR^2(\dot{p} + \dot{\Omega}\cos\gamma)\bar{e}_x + \left(\frac{1}{4}mR^2 + m\ell^2\right)\dot{\Omega}\sin\gamma\,\bar{e}_y$$

$$+ m\sin\gamma\left[\left(\ell^2 - \frac{1}{4}R^2\right)\Omega^2\cos\gamma - \frac{1}{2}R^2 p\Omega\right]\bar{e}_z.$$

Calculations: Substitution of the expressions for $\Sigma\bar{M}_{\mathcal{O}}$ and $\dot{\bar{H}}_{\mathcal{O}}$ into Eq. (17.12), and identification of its x and y components, give $\dot{\Omega} = \dot{p} = 0$. Consequently, Ω and p must both be constant if γ is constant. Identification of the z component gives

$$-mg\ell\sin\gamma = m\sin\gamma\left[\left(\ell^2 - \frac{1}{4}R^2\right)\Omega^2\cos\gamma - \frac{1}{2}R^2 p\Omega\right] \quad\Leftrightarrow$$

$$\left(\ell^2 - \frac{1}{4}R^2\right)\Omega^2\cos\gamma - \frac{1}{2}R^2 p\Omega + g\ell = 0. \tag{17.14}$$

Finally, we can solve Eq. (17.14) for the constant Ω. Particularly, for the case $\gamma = \pi/2$, we observe that the quadratic form disappears, so that Eq. (17.14) can be simplified to

$$\Omega(\gamma = \pi/2) = \frac{2g\ell}{R^2 p}. \qquad\qquad \square$$

The example above shows that one *possible* solution for the motion of the mechanical system in Fig. 17.6 is that the gyroscope (the disc) rotates with a constant angular velocity Ω about a vertical line, while the angle γ remains constant; this satisfies the equations of motion. Experiments show that the system can indeed act this way in practice, provided that the spin is great enough for Eq. (17.14) to have real roots. A rigid body with a spin vector that rotates at a constant angular velocity $\bar{\Omega}$ is said to perform steady-state *precession*.

Appendix

A

Selected mathematics

A.1 Geometry

The *angle* between two straight lines is the rotation required for these lines to align. Consider a circle with radius R, and an arc with length b on this circle. Then the angle θ between the two radii intersecting the edges of this arc is (Fig. A.1)

$$\theta = \frac{b}{R},\qquad\text{(A.1)}$$

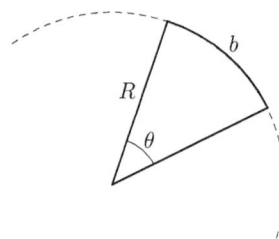

Figure A.1: Geometry for the definition of an angle in the unit of radians.

in the unit of radians (rad), which is the unit for angles in both the SI and the USC system.

For a right triangle with an angle θ, the hypotenuse c, the adjacent side b, and the opposite side a (Fig. A.2), we have

$$\sin\theta = \frac{a}{c},\qquad\text{(A.2a)}$$

$$\cos\theta = \frac{b}{c},\qquad\text{(A.2b)}$$

$$\tan\theta = \frac{\sin\theta}{\cos\theta} = \frac{a}{b},\qquad\text{(A.2c)}$$

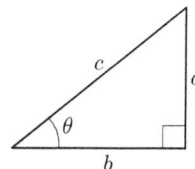

Figure A.2: The geometry for the definition of trigonometric functions.

and the following trigonometric identities hold

$$\sin^2\theta + \cos^2\theta = 1,\qquad\text{(A.3a)}$$

$$\sin(\theta \pm \varphi) = \sin\theta\cos\varphi \pm \cos\theta\sin\varphi,\qquad\text{(A.3b)}$$

$$\cos(\theta \pm \varphi) = \cos\theta\cos\varphi \mp \sin\theta\sin\varphi.\qquad\text{(A.3c)}$$

For a triangle with sides a, b and c, and the angles α, β and γ opposite to those sides (Fig. A.3), the *Law of sines* states that

$$\frac{\sin\alpha}{a} = \frac{\sin\beta}{b} = \frac{\sin\gamma}{c},\qquad\text{(A.4)}$$

while the *Law of cosines* asserts that

$$c^2 = a^2 + b^2 - 2ab\cos\gamma.\qquad\text{(A.5)}$$

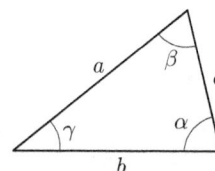

Figure A.3: Geometry for the Laws of sines and cosines.

For two constants A and B, and an angle θ, we have

$$A\cos\theta + B\sin\theta = X\sin(\theta + \psi), \qquad \text{(A.6)}$$

where the *amplitude* X and the *phase angle* ψ are given by

$$X = \sqrt{A^2 + B^2}, \qquad \psi = \begin{cases} \arctan\frac{B}{A}, & A > 0 \\ \arctan\frac{B}{A} + \pi\,\mathrm{sgn}(B), & A < 0 \\ \frac{\pi}{2}\,\mathrm{sgn}(B), & A = 0, \end{cases} \qquad \text{(A.7)}$$

and where $\mathrm{sgn}(\cdot)$ denotes the sign function.

A.2   Vectors

Geometric vectors

A *vector* can be represented geometrically by a directed line segment, which is drawn as an arrow in a plane or in space. A vector pointing out of the paper is drawn using the symbol $\odot$ whereas a vector pointing into the paper is drawn as $\otimes$. Herein, vector quantities are denoted by a bar over the variable name, for instance $\bar{u}$.

The *magnitude* of a vector is denoted by $|\bar{u}|$, being the length of the line segment that represents the vector (Fig. A.4a). Two vectors, $\bar{u}$ and $\bar{w}$, are said to be equal, $\bar{u} = \bar{w}$, if their magnitudes and directions are equal, regardless of their positions in space (Fig. A.4b).

A vector $\overline{\mathcal{AB}}$ is formed by a line segment connecting two points $\mathcal{A}$ and $\mathcal{B}$ (Fig. A.4c). We also introduce the *zero vector* $\bar{0}$, whose magnitude is zero, and whose direction is undefined.

The negated vector $-\bar{u}$ of a vector $\bar{u}$ is a vector with same magnitude as $\bar{u}$, but with the reverse direction (Fig. A.4d).

The *Parallelogram law* states that if the initial point of $\bar{w}$ is placed at the terminal point of $\bar{u}$, then $\bar{u} + \bar{w}$ is the vector from the initial point of $\bar{u}$ to the terminal point of $\bar{w}$ (Fig. A.4e). This operation is called vector addition. Vector subtraction is defined as $\bar{u} - \bar{w} \equiv \bar{u} + (-\bar{w})$.

When a vector $\bar{u}$ is multiplied by a real number c, a new vector $c\bar{u}$ is created. If $c > 0$, then $\bar{u}$ and $c\bar{u}$ have the same direction. However, if $c < 0$, then the vectors $\bar{u}$ and $c\bar{u}$ have opposite directions. It holds that $c\bar{u}$ is $|c|$ times longer than $\bar{u}$, and that $0\bar{u} = \bar{0}$.

The following operations of vector arithmetic hold in two- and three-dimensional space:

$$\bar{u} + \bar{w} = \bar{w} + \bar{u}, \qquad \text{(A.8a)}$$

$$c(d\bar{u}) = (cd)\bar{u}, \qquad \text{(A.8b)}$$

$$c(\bar{u} + \bar{w}) = c\bar{u} + c\bar{w}, \qquad \text{(A.8c)}$$

$$(c + d)\bar{u} = c\bar{u} + d\bar{u}. \qquad \text{(A.8d)}$$

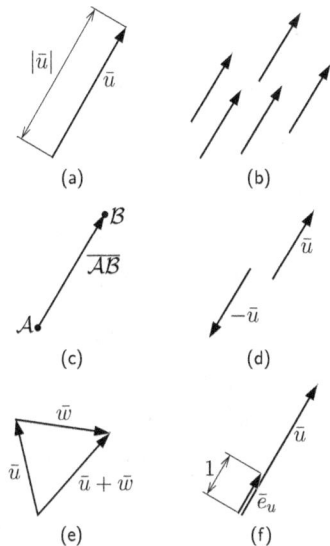

Figure A.4: (a) Vector $\bar{u}$ with magnitude $|\bar{u}|$. (b) Equal vectors. (c) Vector connecting two points. (d) Negation reverses the direction of a vector. (e) Vector addition with the Parallelogram law. (f) The direction vector $\bar{e}_u$ of $\bar{u}$ has the same direction as $\bar{u}$, but the magnitude 1.

Here, c and d denote arbitrary real numbers.

A vector with magnitude 1 is called a *unit vector*. An arbitrary vector $\bar{u} \neq \bar{0}$ has a *direction vector* $\bar{e}_u$, which is a unit vector parallel to $\bar{u}$ (Fig. A.4f). Therefore, we write

$$\bar{u} = u\bar{e}_u \quad \Leftrightarrow \quad \bar{e}_u = \frac{\bar{u}}{u}, \tag{A.9}$$

where $u \neq 0$ is a scalar, that is a real number. This scalar is allowed to be positive or negative.

Vectors in orthogonal coordinate systems

Consider an orthogonal, right-hand oriented coordinate system with origin $\mathcal{O}$ and the coordinates x, y and z. A coordinate system is said to be *orthogonal* if its axes are perpendicular to each other. The right-hand rule is used to ascertain that a coordinate system is right-hand oriented (Fig. A.5).

Every coordinate axis x, y and z define a direction vector $\bar{e}_x$, $\bar{e}_y$ and $\bar{e}_z$, respectively, in the positive coordinate directions (Fig. A.6a). These vectors $\bar{e}_x$, $\bar{e}_y$ and $\bar{e}_z$ form an orthonormal basis, so that an arbitrary vector $\bar{u}$ can be uniquely represented as

$$\bar{u} = u_x\bar{e}_x + u_y\bar{e}_y + u_z\bar{e}_z, \tag{A.10}$$

where the scalars u_x, u_y and u_z are the *components* of vector $\bar{u}$ (Fig. A.6b). Whenever it is convenient, an equivalent notation is used, where vectors are represented as column matrices:

$$u_x\bar{e}_x + u_y\bar{e}_y + u_z\bar{e}_z = \begin{bmatrix} u_x \\ u_y \\ u_z \end{bmatrix}.$$

The representation of a vector in an orthonormal basis is unique. Owing to this property, it holds that

$$\bar{u} = \bar{w} \quad \Leftrightarrow \quad \begin{cases} u_x = w_x \\ u_y = w_y \\ u_z = w_z. \end{cases} \tag{A.11}$$

Thus, a vector equation can be written as a system of scalar equations with real coefficients and real variables.

Scalar product

The *scalar product*[30] between two vectors $\bar{u} = u_x\bar{e}_x + u_y\bar{e}_y + u_z\bar{e}_z$ and $\bar{w} = w_x\bar{e}_x + w_y\bar{e}_y + w_z\bar{e}_z$ is defined as

$$\bar{u} \cdot \bar{w} \equiv |\bar{u}||\bar{w}| \cos\varphi, \tag{A.12}$$

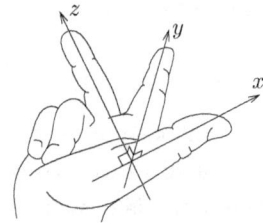

Figure A.5: The right-hand rule: When holding the first three fingers of the right hand perpendicular to each other, they will point in the x, y and z the directions respectively.

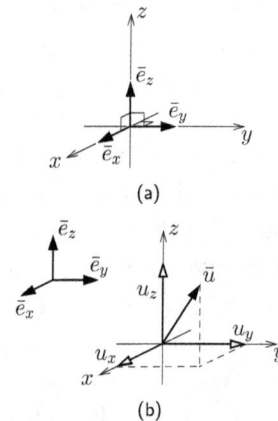

Figure A.6: (a) A right-hand, orthogonal coordinate system, with orthonormal basis $\bar{e}_x$, $\bar{e}_y$ and $\bar{e}_z$. (b) A vector $\bar{u}$ with its three components $u_x\bar{e}_x$, $u_y\bar{e}_y$ and $u_z\bar{e}_z$, which are drawn with open arrowheads.

[30] Also called *dot product*.

where φ is the angle between $\bar{u}$ and $\bar{w}$. Its result is a scalar, and it can be shown that

$$\bar{u} \cdot \bar{w} = u_x w_x + u_y w_y + u_z w_z. \tag{A.13}$$

It follows from Eq. (A.12), that if $\bar{u}, \bar{w} \neq 0$, and the angle between them is $\varphi = \pi/2$, then the scalar product becomes zero

$$\bar{u} \perp \bar{w} \quad \Leftrightarrow \quad \bar{u} \cdot \bar{w} = 0. \tag{A.14}$$

Moreover, since $\cos 0 = 1$, Eq. (A.12) implies that $\bar{u} \cdot \bar{u} = |\bar{u}|^2$. From this property of the scalar product, we obtain an expression for the magnitude of a vector

$$|\bar{u}| = \sqrt{\bar{u} \cdot \bar{u}} = \sqrt{u_x^2 + u_y^2 + u_z^2}. \tag{A.15}$$

Some useful algebraic rules of the scalar product are:

$$\bar{u} \cdot \bar{w} = \bar{w} \cdot \bar{u}, \tag{A.16a}$$

$$\bar{u} \cdot (\bar{v} + \bar{w}) = \bar{u} \cdot \bar{v} + \bar{u} \cdot \bar{w}, \tag{A.16b}$$

$$c(\bar{u} \cdot \bar{w}) = (c\bar{u}) \cdot \bar{w}, \tag{A.16c}$$

where c is a scalar. From these rules it follows that

$$\bar{u} \cdot \bar{e}_x = u_x(\bar{e}_x \cdot \bar{e}_x) + u_y(\bar{e}_y \cdot \bar{e}_x) + u_z(\bar{e}_z \cdot \bar{e}_x)$$
$$= u_x 1 + u_y 0 + u_z 0 = u_x.$$

This result can be generalized to an arbitrary axis λ with direction vector $\bar{e}_\lambda$. We have that $\bar{u} \cdot \bar{e}_\lambda$ is the component of $\bar{u}$ in the λ direction. The scalar product with the unit vector $\bar{e}_\lambda$ can be interpreted as an orthogonal projection onto the λ axis:

$$u_\lambda = \bar{u} \cdot \bar{e}_\lambda = |\bar{u}| \cos \varphi, \tag{A.17}$$

where φ is the angle between $\bar{u}$ and $\bar{e}_\lambda$ (Fig. A.7).

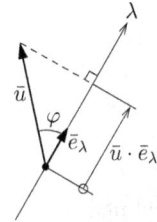

Figure A.7: Projection of a vector onto an arbitrary axis λ through scalar multiplication with a direction vector.

Cross product

The *cross product* $\bar{u} \times \bar{w}$ between two vectors is defined using determinant notation:

$$\bar{u} \times \bar{w} \equiv \begin{vmatrix} \bar{e}_x & \bar{e}_y & \bar{e}_z \\ u_x & u_y & u_z \\ w_x & w_y & w_z \end{vmatrix} =$$

$$= (u_y w_z - u_z w_y)\bar{e}_x + (u_z w_x - u_x w_z)\bar{e}_y + (u_x w_y - u_y w_x)\bar{e}_z. \tag{A.18}$$

The result of a cross product is a vector with magnitude

$$|\bar{u} \times \bar{w}| = |\bar{u}||\bar{w}| \sin \varphi, \tag{A.19}$$

where φ is again the angle between $\bar{u}$ and $\bar{w}$. This magnitude is equal to the area of the parallelogram spanned by $\bar{u}$ and $\bar{w}$. Furthermore, $\bar{u} \times \bar{w}$ is perpendicular to both $\bar{u}$ and $\bar{w}$, and its orientation follows the right-hand rule (Fig. A.8). According to Eq. (A.19), if $\bar{u}, \bar{w} \neq \bar{0}$ and $\varphi = 0$ or $\varphi = \pi$, then the cross product becomes $\bar{0}$:

$$\bar{u} \parallel \bar{w} \quad \Leftrightarrow \quad \bar{u} \times \bar{w} = \bar{0}. \tag{A.20}$$

Some useful algebraic rules of the cross product are

$$\bar{u} \times \bar{w} = -(\bar{w} \times \bar{u}), \tag{A.21a}$$

$$\bar{u} \times (\bar{v} + \bar{w}) = \bar{u} \times \bar{v} + \bar{u} \times \bar{w}, \tag{A.21b}$$

$$c(\bar{u} \times \bar{w}) = (c\bar{u}) \times \bar{w} = \bar{u} \times (c\bar{w}), \tag{A.21c}$$

$$\bar{u} \times \bar{u} = \bar{0}, \tag{A.21d}$$

where c is a scalar. In addition, there are vector identities that include both scalar and cross products:

$$\bar{u} \times (\bar{v} \times \bar{w}) = (\bar{u} \cdot \bar{w})\bar{v} - (\bar{u} \cdot \bar{v})\bar{w}, \tag{A.22a}$$

$$\bar{u} \cdot (\bar{v} \times \bar{w}) = \bar{w} \cdot (\bar{u} \times \bar{v}) = \bar{v} \cdot (\bar{w} \times \bar{u}). \tag{A.22b}$$

Vector-valued functions

If the value of a vector $\bar{u}$ depends on a variable t, then $\bar{u}(t)$ is a *vector-valued function*. Using a space-fixed coordinate system xyz, we write

$$\bar{u}(t) = u_x(t)\bar{e}_x + u_y(t)\bar{e}_y + u_z(t)\bar{e}_z, \tag{A.23}$$

where $u_x(t)$, $u_y(t)$ and $u_z(t)$ are scalar functions, and where $\bar{e}_x$, $\bar{e}_y$ and $\bar{e}_z$ are constant unit vectors.[31]

The derivative of a vector-valued function $\bar{u}(t)$ with constant basis vectors is defined as

$$\frac{d\bar{u}}{dt} \equiv \lim_{\Delta t \to 0} \frac{\bar{u}(t + \Delta t) - \bar{u}(t)}{\Delta t} = \frac{du_x}{dt}\bar{e}_x + \frac{du_y}{dt}\bar{e}_y + \frac{du_z}{dt}\bar{e}_z. \tag{A.24}$$

The Product rule holds when expressions containing vector-valued functions are differentiated w.r.t. t:

$$\frac{d}{dt}(c\bar{u}) = \frac{dc}{dt}\bar{u} + c\frac{d\bar{u}}{dt}, \tag{A.25a}$$

$$\frac{d}{dt}(\bar{u} \cdot \bar{w}) = \frac{d\bar{u}}{dt} \cdot \bar{w} + \bar{u} \cdot \frac{d\bar{w}}{dt}, \tag{A.25b}$$

$$\frac{d}{dt}(\bar{u} \times \bar{w}) = \frac{d\bar{u}}{dt} \times \bar{w} + \bar{u} \times \frac{d\bar{w}}{dt}, \tag{A.25c}$$

where c, $\bar{u}$ and $\bar{w}$ are functions of t. Moreover, the integral of a vector-valued function $\bar{u}(t)$ is

$$\int_{t_1}^{t_2} \bar{u}\,dt = \int_{t_1}^{t_2} u_x\,dt\,\bar{e}_x + \int_{t_1}^{t_2} u_y\,dt\,\bar{e}_y + \int_{t_1}^{t_2} u_z\,dt\,\bar{e}_z. \tag{A.26}$$

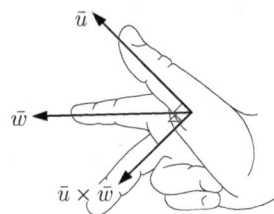

Figure A.8: The right-hand rule gives the direction of the result vector of the cross product.

[31] In general, vector-valued functions can depend on several variables, and furthermore, other ranges than $\mathbb{R}^3$.

For a *moving* coordinate system xyz, the basis vectors $\bar{e}_x(t)$, $\bar{e}_y(t)$ and $\bar{e}_z(t)$ depend on the variable t. Even though their directions vary, these unit vectors remain orthogonal. Therefore, the vector-valued function $\bar{u}(t)$ is written as

$$\bar{u}(t) = u_x(t)\bar{e}_x(t) + u_y(t)\bar{e}_y(t) + u_z(t)\bar{e}_z(t). \tag{A.27}$$

The Product rule of differentiation holds, so that

$$\frac{d\bar{u}}{dt} = \frac{d}{dt}\left(u_x\bar{e}_x + u_y\bar{e}_y + u_z\bar{e}_z\right) = \{\text{Eq. (A.25a)}\}$$

$$= \frac{du_x}{dt}\bar{e}_x + u_x\frac{d\bar{e}_x}{dt} + \frac{du_y}{dt}\bar{e}_y + u_y\frac{d\bar{e}_y}{dt} + \frac{du_z}{dt}\bar{e}_z + u_z\frac{d\bar{e}_z}{dt}. \tag{A.28}$$

A.3   Differentials

For a function $y(t)$, the symbol dy/dt denotes the derivative of y w.r.t. t. This notation should *not* be regarded as a quotient between a numerator dy and a denominator dt, since such a ratio would be $0/0$, which is undefined. Instead, dy/dt should be regarded as a *symbol* for the derivative. However, we can introduce an independent variable dt called the *differential* of t. Moreover, we regard dy as a dependent variable called the *differential* of y. Both these variables, dy and dt, are allowed to have finite values.

Definition A.1 (Differential). If $y(t)$ is a differentiable function, the differential of y is

$$dy \equiv \frac{dy}{dt}dt, \tag{A.29}$$

where dt is an independent variable called the differential of t.

As usual, dy/dt in Eq. (A.29) denotes the derivative of y. We conclude that $dy = dy(t, dt)$ is a function of two variables t and dt (Fig. A.9).

For a given function $f(t)$, Def. (A.29) gives

$$dy = f(t)dt \quad \Leftrightarrow \quad \frac{dy}{dt} = f(t). \tag{A.30}$$

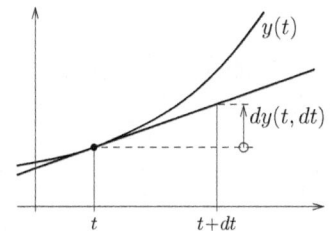

Figure A.9: The relation between the function $y(t)$ and its differential $dy = dy(t, dt)$.

Thus, a differential expression offers an alternative notation for the derivative. Since both sides of Eq. (A.29) evaluate to real numbers, the differential notation opens new possibilities: Algebraic operations that are permissible for scalar equations, are also permissible for equations that contain differentials.

The following two theorems demonstrate how differentials can be rewritten into integral equations:

138 LECTURES ON ENGINEERING MECHANICS

Theorem A.2 (Separable differential equations). If $y(t)$ is a differentiable function, and if $f(t)$ and $g(y)$ are given functions, it holds that

$$g[y(t)]dy = f(t)dt \quad \Leftrightarrow \quad \int_{y(t_1)}^{y(t_2)} g(y)\mathrm{d}y = \int_{t_1}^{t_2} f(t)\mathrm{d}t. \qquad \text{(A.31)}$$

Theorem A.3. If $x(t)$ and $y(t)$ are differentiable functions, and if $g(y)$ and $h(x)$ are given functions, then it holds that

$$g[y(t)]dy = h[x(t)]dx \quad \Leftrightarrow \quad \int_{y(t_1)}^{y(t_2)} g(y)\mathrm{d}y = \int_{x(t_1)}^{x(t_2)} h(x)\mathrm{d}x. \qquad \text{(A.32)}$$

The concept of differentials can be extended to vector-valued functions. If $\bar{u}(t) = u_x \bar{e}_x + u_y \bar{e}_y + u_z \bar{e}_z$ is a differentiable, vector-valued function, then its differential is

$$d\bar{u} \equiv du_x \bar{e}_x + du_y \bar{e}_y + du_z \bar{e}_z. \qquad \text{(A.33)}$$

It follows from Eq. (A.29) that

$$d\bar{u} = \frac{du_x}{dt}dt\bar{e}_x + \frac{du_y}{dt}dt\bar{e}_y + \frac{du_y}{dt}dt\bar{e}_z = \frac{d\bar{u}}{dt}dt, \qquad \text{(A.34)}$$

and therefore, that

$$d\bar{u} = \bar{w}(t)dt \quad \Leftrightarrow \quad \frac{d\bar{u}}{dt} = \bar{w}(t). \qquad \text{(A.35)}$$

Integration from t_1 and t_2 on the right-hand side of the implication in Eq. (A.35) gives

$$d\bar{u} = \bar{w}(t)dt \quad \Leftrightarrow \quad \bar{u}(t_2) - \bar{u}(t_1) = \int_{t_1}^{t_2} \bar{w}(t)\mathrm{d}t. \qquad \text{(A.36)}$$

A.4 Integrals

The reader is assumed to be acquainted with proper integrals. Here, we formulate some of their fundamental properties.

Theorem A.4 (Fundamental theorem of calculus). If the function f is continuous on the interval $[a, b]$, then it holds that

$$\frac{\mathrm{d}}{\mathrm{d}x} \int_a^x f(\xi)\mathrm{d}\xi = f(x), \qquad a \le x \le b. \qquad \text{(A.37)}$$

A useful technique for simplifying integral expressions is the method of substitution:

Theorem A.5 (Integration by substitution)**.** If the function g is differentiable on the interval $[a, b]$, and if the function f is continuous in the range of g on the domain $[a, b]$, then it holds that

$$\int_a^b f[g(t)]\frac{\mathrm{d}g}{\mathrm{d}t}\mathrm{d}t = \int_{g(a)}^{g(b)} f(u)\mathrm{d}u, \qquad (\text{A.38})$$

where $u = g(t)$ is the *substitution*, and $du = \frac{\mathrm{d}g}{\mathrm{d}t}dt$.

To avoid that different notations are used for same quantity, one can reformulate Eq. (A.38) as

$$\int_a^b f[g(t)]\frac{\mathrm{d}g}{\mathrm{d}t}\mathrm{d}t = \int_{g(a)}^{g(b)} f(g)\mathrm{d}g, \qquad (\text{A.39})$$

where the substitution is $g = g(t)$. Thus, g takes the alternate roles of integrand and integration variable.

The principle of *localization* is often useful for simplifying integral equations. Briefly, it states that if an integral vanishes for all integration domains, then the integrand must be identically zero.

Theorem A.6 (Localization)**.** If the function $f(\bar{r})$ is continuous in an open domain Ω_0, it holds that

$$\int_{\Omega} f(\bar{r})\mathrm{d}V = 0, \ \forall \Omega \subset \Omega_0 \quad \Leftrightarrow \quad f(\bar{r}) = 0, \ \forall \bar{r} \in \Omega_0. \qquad (\text{A.40})$$

This principle of localization can be shown in one and two and three dimensions.

B
Quantity, unit and dimension

A *quantity* is a measurable property of an object or a phenomenon. Every quantity possesses a *physical dimension* and a *magnitude*. The *dimension* of the quantity refers to the type of quantity considered, for instance, length, time, velocity, mass or force. The magnitude of a quantity is its relative size compared to some reference quantity with the same dimension.

The fundamental dimensions in mechanics are time (T), length (L) and mass (M). From these dimensions T, L and M, we can derive other dimensions. Since velocity is defined as a distance (L) per unit time (T), the dimension of velocity is L/T. In the same way, the dimension of acceleration is L/T^2. According to Eq. (A.1), an angle is the ratio between the length (L) and the radius (L) of a circular arc. Therefore, the dimension of an angle is L/L = 1. A quantity is said to be *dimensionless* when it has the dimension 1.

A *unit* is a well-defined quantity used as a reference when specifying the magnitude of other quantities of same dimension. The *International system of units* defines a set of base units from physical phenomena through universal constants.

International system of units

In the International system of units (SI), the *second*, the *meter*, and the *kilogram* are used as the base units of time (T), length (L), and mass (M), respectively.

Derived units can be formed from products or quotients of previously defined units. For instance, we can form the unit meter per second (m/s), which has the physical dimension L/T. Thus, m/s can be used to describe velocity. In the SI system of units, the units of second, meter and kilogram are defined through universal constants, which themselves have derived units.[32]

[32] Bureau International des Poids et Mesures. The International System of Units (SI), 2019

Definition B.1 (Universal constants). The frequency of transition between the two hyperfine energy levels of the Cesium-133 isotope, being in its ground state at absolute zero, is a constant

$$\Delta\nu_{\text{Cs}} \equiv 9\,192\,631\,770\,\frac{1}{\text{s}}. \tag{B.1}$$

The *speed of light* in vacuum is a constant

$$c \equiv 299\,792\,458\,\frac{\text{m}}{\text{s}}. \tag{B.2}$$

The *Planck constant* is defined as

$$h \equiv 6.62607015 \cdot 10^{-34}\,\frac{\text{kg}\cdot\text{m}^2}{\text{s}}. \tag{B.3}$$

The constant $\Delta\nu_{\text{Cs}}$ has been chosen as a reference, since it can be measured with high accuracy and precision using an atomic clock. According to Eq. (B.1), the unit of second (s) is given by

$$1\,\text{s} = 9\,192\,631\,770\,\frac{1}{\Delta\nu_{\text{Cs}}}, \tag{B.4}$$

corresponding to the duration of $9\,192\,631\,770$ periods of this Cesium-133 radiation. Moreover, from Eq. (B.2) we obtain the expression for a meter (m)

$$1\,\text{m} = \left(\frac{1}{299\,792\,458}\,\text{s}\right)c, \tag{B.5}$$

so that the meter is the distance that light travels in vacuum during $1/299\,792\,458$ second. The unit of kilogram (kg) is determined through Eqs. (B.1), (B.2) and (B.3) to be

$$1\,\text{kg} = \left(\frac{299\,792\,458^2}{6.62607015 \cdot 10^{-34} \cdot 9\,192\,631\,770}\right)\frac{h\Delta\nu_{\text{Cs}}}{c^2}. \tag{B.6}$$

However, this latter equation does not have any obvious physical interpretation in Classical Mechanics.

There are also dimensionless units with the physical dimension 1, for instance the per cent (%), and the units of radians (rad) or degrees (°) for angles. These units are mathematically defined without reference to any physical phenomenon.

Prefixes can be prepended to SI units, where they denote a multiple or a fraction of a unit. For instance, a microsecond (μs) denotes one part of a million of a second, where micro- (μ) is the prefix for one part per million. Some of the most common prefixes are listed in Table B.1.

Prefix	Symbol	Factor
tera-	T	10^{12}
giga-	G	10^{9}
mega-	M	10^{6}
kilo-	k	10^{3}
hecto-	h	10^{2}
deci-	d	10^{-1}
centi-	c	10^{-2}
milli-	m	10^{-3}
micro-	μ	10^{-6}
nano-	n	10^{-9}
pico-	p	10^{-12}

Table B.1: Some prefixes used in the SI system.

U.S. Customary Units

U.S. Customary Units (USC) is a system of units, which is mostly used in the United States. The USC system uses the second as the base unit for time (T), as defined in Eq. (B.4). Other USC units are defined as multiples of SI units.

Definition B.2. The *foot* (ft) is exactly

$$1\,\text{ft} \equiv 0.3048\,\text{m}, \tag{B.7}$$

and the *pound-mass* (lb_m) is exactly

$$1\,\text{lb}_\text{m} \equiv 0.453\,592\,37\,\text{kg}, \tag{B.8}$$

by definition.[33]

The USC base unit for length (L) is the foot (ft). The third USC base unit is the *pound-force* (lb_f) defined as the force that acts on $1\,\text{lb}_\text{m}$ in the standard gravity $g_\text{n} \equiv 9.80665\,\text{N/kg}$ of the SI system:

$$1\,\text{lb}_\text{f} \equiv 1\,\text{lb}_\text{m} \cdot g_\text{n} \approx 4.4482\,\text{N}. \tag{B.9}$$

The reader should be aware of the distinction between the reference mass, *i.e.* the pound-mass (lb_m), and the base unit for force, *i.e.* the pound-force (lb_f). The USC unit for mass (M) is the *slug*, which is derived from the base units s, ft and lb_f:

$$1\,\text{slug} \equiv 1\,\frac{\text{lb}_\text{f} \cdot \text{s}^2}{\text{ft}} \approx 14.594\,\text{kg}. \tag{B.10}$$

Additional USC units are derived from the base units.[34]

All physical equations are formulated so that they are valid only for base units. Consequently, it is necessary to use either the base units {s, m, kg, rad} of the SI system, or the base units {s, ft, lb_f, rad} of the USC system. Particularly, degrees must always be converted to radians, and the various units of mass used within the USC system, including lb_m, must be converted to slugs.

Numerical value

The value of a scalar quantity X w.r.t. a unit E is expressed as the product between a *numerical value* n and the unit itself:

$$X = nE, \tag{B.11}$$

where n is a real coefficient that does not affect the dimension of the expression. A vector quantity $\bar{X}$ can be written:

$$\bar{X} = \bar{n}E, \tag{B.12}$$

where $\bar{n}$ is a vector with real components. If the velocity $\bar{v}$ is $5.0\,\text{m/s}$ in the z direction, it is proper to write $\bar{v} = 5.0\bar{e}_z\,\text{m/s}$.

[33] U.S. National Bureau of Standards. *Research Highlights of the National Bureau of Standards.* U.S. Department of Commerce, National Bureau of Standards, 1959

[34] Some units of the USC system:

inch:	$1\,\text{in} \equiv \dfrac{1}{12}\,\text{ft}$
yard:	$1\,\text{yd} \equiv 3\,\text{ft}$
mile:	$1\,\text{mi} \equiv 5280\,\text{ft}$
ounce:	$1\,\text{oz} \equiv \dfrac{1}{16}\,\text{lb}_\text{m}$
pound-mass:	$1\,\text{lb}_\text{m} \approx 0.031081\,\text{slug}$

Algebraic rules for dimensions

We denote the dimension of a quantity by $[\![X]\!]$. For instance, the expression $[\![m]\!] = M$ means that m has the dimension of mass. If both X and Y are quantities, it holds that

$$X = Y \quad \Rightarrow \quad [\![X]\!] = [\![Y]\!]. \tag{B.13}$$

Consequently, the dimension must be the same on both sides of an equation. These algebraic rules for dimensions of quantities hold:

$$[\![nX]\!] = [\![X]\!] \tag{B.14a}$$

$$[\![X + Y]\!] = \begin{cases} [\![X]\!], & \text{if } [\![X]\!] = [\![Y]\!] \\ \text{undefined}, & \text{otherwise} \end{cases} \tag{B.14b}$$

$$[\![XY]\!] = [\![X]\!]\,[\![Y]\!] \tag{B.14c}$$

$$[\![X^n]\!] = [\![X]\!]^n \tag{B.14d}$$

where n is a real number. A dimensional analysis of the Law of force and acceleration, Eq. (1.4), gives

$$[\![\Sigma \bar{F}]\!] = [\![m\bar{a}]\!] = \{\text{Eq. (B.14c)}\} = [\![m]\!]\,[\![\bar{a}]\!] = \frac{ML}{T^2}.$$

Thus, the dimension of force is obtained as a combination of the fundamental dimensions. Since the units themselves are quantities, it is proper to write

$$[\![s]\!] = T, \quad [\![ft]\!] = L, \quad [\![kg]\!] = [\![slug]\!] = M, \quad [\![m/s]\!] = \frac{L}{T}, \quad [\![lb_f]\!] = \frac{ML}{T^2},$$

and so on. For this reason, it is convenient to use the base units to represent dimensions. For instance, the dimension of velocity can be written as $[\![m/s]\!]$.

Dimensional correctness

All physical equations and expressions must be dimensionally correct. This means that it is required that:

- The dimensions must be equal on the different sides of equalities and inequalities.

- The dimensions of all the terms in a sum must be equal.

- The dimension of the argument x of transcendental functions, for instance, $\cos x$, $\ln x$ or e^x, must be $\mathbf{1}$.

To identify errors during problem solving, it is instrumental to check all expressions and equations for dimensional correctness.

C
Tables

A	α	Alpha	N	ν	Nu
B	β	Beta	Ξ	ξ	Xi
Γ	γ	Gamma	O	o	Omicron
Δ	δ	Delta	Π	π	Pi
E	ε	Epsilon	P	ρ, ϱ	Rho
Z	ζ	Zeta	Σ	σ	Sigma
H	η	Eta	T	τ	Tau
Θ	θ	Theta	Υ	υ	Ypsilon
I	ι	Iota	Φ	ϕ, φ	Phi
K	κ	Kappa	X	χ	Chi
Λ	λ	Lambda	Ψ	ψ	Psi
M	μ	Mu	Ω	ω	Omega

Table C.1: Letters of the Greek alphabet.

Table C.2: The diagonal elements $I_{\mathcal{G}xx}$, $I_{\mathcal{G}yy}$ and $I_{\mathcal{G}zz}$ and all nonzero off-diagonal elements of the inertia matrix w.r.t. the center of mass $\mathcal{G}$ for three-dimensional bodies and shells with uniformly distributed mass m.

Body		Moments and products of inertia
Sphere		$I_{\mathcal{G}xx} = I_{\mathcal{G}yy} = I_{\mathcal{G}zz} = \frac{2}{5}mr^2$
Rectangular parallelepiped		$I_{\mathcal{G}xx} = \frac{1}{12}m(b^2 + c^2)$ $I_{\mathcal{G}yy} = \frac{1}{12}m(a^2 + c^2)$ $I_{\mathcal{G}zz} = \frac{1}{12}m(a^2 + b^2)$

Cont.

Cont.

Body	Moments and products of inertia
Right triangular prism	$I_{\mathcal{G}xx} = \frac{1}{18}mb^2 + \frac{1}{12}mc^2$ $I_{\mathcal{G}yy} = \frac{1}{18}ma^2 + \frac{1}{12}mc^2$ $I_{\mathcal{G}zz} = \frac{1}{18}m(a^2 + b^2)$ $I_{\mathcal{G}xy} = \frac{1}{36}mab$
Circular cylinder	$I_{\mathcal{G}xx} = I_{\mathcal{G}yy} = \frac{1}{4}mr^2 + \frac{1}{12}m\ell^2$ $I_{\mathcal{G}zz} = \frac{1}{2}mr^2$
Semicylinder	$I_{\mathcal{G}xx} = \left(\frac{1}{4} - \frac{16}{9\pi^2}\right)mr^2 + \frac{1}{12}m\ell^2$ $I_{\mathcal{G}yy} = \frac{1}{4}mr^2 + \frac{1}{12}m\ell^2$ $I_{\mathcal{G}zz} = \left(\frac{1}{2} - \frac{16}{9\pi^2}\right)mr^2$
Cone	$I_{\mathcal{G}xx} = I_{\mathcal{G}yy} = \frac{3}{20}mr^2 + \frac{3}{80}mh^2$ $I_{\mathcal{G}zz} = \frac{3}{10}mr^2$
Spherical shell	$I_{\mathcal{G}xx} = I_{\mathcal{G}yy} = I_{\mathcal{G}zz} = \frac{2}{3}mr^2$
Cylindrical shell	$I_{\mathcal{G}xx} = I_{\mathcal{G}yy} = \frac{1}{2}mr^2 + \frac{1}{12}m\ell^2$ $I_{\mathcal{G}zz} = mr^2$
Half cylindrical shell	$I_{\mathcal{G}xx} = \left(\frac{1}{2} - \frac{4}{\pi^2}\right)mr^2 + \frac{1}{12}m\ell^2$ $I_{\mathcal{G}yy} = \frac{1}{2}mr^2 + \frac{1}{12}m\ell^2$ $I_{\mathcal{G}zz} = \left(1 - \frac{4}{\pi^2}\right)mr^2$

Index

acceleration, 3
  angular, 45
  def. of, 43
  in arc coordinates, 49
  in natural basis, 49
  in polar coordinates, 47
  in rectangular coordinates, 44
  instantaneous, 42
  of gravity, 5
  relative, 86, 115
addition of vectors, 133
altitude coordinate, 60
amplitude
  def. of, 133
  of harmonic motion, 74, 79
angle
  def. of, 132
  polar, 83
angular acceleration
  for general motion, 112
  for planar motion, 84
  of polar coordinates, 45
angular frequency
  natural, 74
  resonance, 79
angular impulse, 107
angular impulse–angular momentum
    relation
  for body, 106
  for particle, 65
  for rigid body, 108
  for system of particles, 68
angular momentum
  and kinetic energy, 123
  for general motion, 121
  for planar motion, 95
  of body, 90

  of particle, 64
  of system of particles, 66
  transfer theorem of, 92
  w.r.t. body- and space-fixed point,
    97, 123
  w.r.t. center of mass, 96, 122
angular velocity
  for general motion, 111
  for planar motion, 83
  of coordinate system, 111
  of polar coordinates, 45
  relative, 115, 116
arc coordinate, 47
arc length, 47
Archimedes' principle, 35
axis of rotation, 83, 88

basis vector
  natural, 47
  polar, 45
  rectangular, 43
beam, 27
belt friction, 39
bending moment, 28
body, 2
  composite, 22
  rigid, def. of, 2
body force, 7, 23
body-fixed point, 86
body-fixed vector, 114
buoyancy force, 34

c (speed of light), 141
center of curvature, 48
center of gravity, 16, 21, 23
center of loading, 25
center of mass, 22

central force, 5
centroid, 22
circular motion, 47
coefficient of kinetic friction, 37
coefficient of restitution, 71
coefficient of static friction, 37
collision, 69, 108
component
  of force, 7
  of vector, 134
composite body, 22
conservation
  of angular momentum, 69
  of momentum, 68, 69
conservative force, 60
constant
  gravitational, 5
  local gravity, 5, 16
  Planck, 141
  spring, 6
  universal, 140
constraint, 49
constraint couple, 16
constraint force, 16, 59
contact force, 5, 7
coordinate
  altitude, 60
  arc, 47
  depth, 34
  height, 60
  natural, 47
  polar, 45
  rectangular, 43
coordinate system
  natural, 47
  orthogonal, 134
  polar, 45

rectangular, 43
right-hand, 134
rotating, 113
Coriolis equation, 112, 114
Coulomb friction, 37
couple
  constraint, 16
  def. of, 9
  internal, 27, 28
  power of, 99
  section, 27
critically damped, 76
cross product, 135
cross-section, 28
curvilinear motion, 43

damped oscillator, 75
damper, 75
damping, 75
damping coefficient, 75
damping force, 75
damping ratio, 76
density, 21
  line, 23
  surface, 23, 94
depth coordinate, 34
derivative
  of basis vector, 113
  of polar basis vector, 46
  of vector-valued function, 136
differential, 137
differential equation
  second-order, 76
  separable, 138
dimension, 140
dimensional correctness, 143
dimensionless, 140
direct central impact, 70
direction vector, 134
distributed load, 25
dot product, 134
dry friction, 37
dynamic unbalance, 124

elastic energy, 61
elastic impact, 71
element
  line, 23
  mass, 21, 94
  surface, 23, 94

volume, 21
energy
  elastic, 61
  kinetic, 59, 102, 123
  mechanical, 60
  potential, 60
equilibrium
  condition for, 14
  force, 14
  moment, 14
  of beam, 29
  static, 14
equilibrium position, 74
Euler's first law, 91
Euler's laws of motion, 90, 91
Euler's second law, 92
external force, 67

fictitious force, 53
field of gravity, 23
fixed-axis rotation, 88, 124
flat rigid body, 94
fluid, 30
fluid statics, 30
foot (ft), 142
force, 3, 7
  body, 7, 23
  buoyancy, 34
  central, 5
  conservative, 60
  constraint, 16, 59
  contact, 5, 7
  damping, 75
  external, 7, 67
  fictitious, 53
  friction, 36
  gravitational, 5
  internal, 27, 28, 67
  normal, 28, 36
  of action and reaction, 51
  of gravity, 16
  section, 27
  shear, 28
  spring, 6
  tensile, 18
force component, 7
force equation, 91, 92, 118
force equilibrium, 14
force of gravity, 23
force pair, 9

force sum, 10
force vector, 7
force-couple system
  def. of, 10
  planar, 12
  reduced, 11
forced harmonic oscillator, 76
frame of reference, 52
free harmonic oscillators, 73
free-body diagram
  construction of, 15
  for impulse system, 109
  of beam, 29
  of multi-body system, 19
friction
  belt, 39
  Coulomb, 37
  dry, 37
  kinetic, 37
  static, 37
friction force, 36
frictionless surface, 36
function
  harmonic, 75
  transcendental, 143
  vector-valued, 136, 138
fundamental theorem of calculus, 138

g (local gravity constant), 5
G_{g} (gravitational constant), 5
gas, 30
geometry, 132
gravitation, 5
gravitational constant, 5
gravitational force, 5
gravity, 16, 23
Greek alphabet, 144
gyrodynamics, 127
gyroscope, 127

h (Planck constant), 141
harmonic function, 75
harmonic motion, 74
harmonic oscillator, 73
  forced, 76
harmonic oscillators
  free, 73
height coordinate, 60, 104
hertz (Hz), 74
homogeneous, 22

homogeneous solution, 77
horsepower (hp), 57

impact
    between particles, 69
    direct central, 70
    elastic, 71
    oblique, 71
    plastic, 71
impact impulse, 71
impact model, 70
impulse, 64, 72
impulse couple, 107
impulse pair, 107
impulse system, 108
impulse–momentum relation
    for body, 106
    for particle, 63
    for rigid body, 107
    for system of particles, 67
inertia, 4, 51, 94
inertia matrix
    def. of, 118
    table of, 144
    transfer of, 121
inertia tensor, 118
inertial frame of reference, 52
inertial system, 4, 52
initial condition, 74
instantaneous acceleration, 42
instantaneous center, 86
instantaneous impact model
    for particle, 70
    for rigid body, 108
instantaneous velocity, 42
integral, 138
integration by substitution, 139
interaction, 3
internal couple, 27, 28
internal force, 27, 28, 67
international system of units, 140

joint, 17
joule (J), 57

kilogram (kg), 140
kinematic constraint, 49
kinematics
    planar, 42, 82
    three-dimensional, 111

kinetic energy
    for general motion, 123
    of body, 102
    of particle, 59
    of rigid body, 102
kinetic friction, 37
kinetics
    of particles, 51
    planar, 90
    three-dimensional, 118
kinetics problem, 53

law
    Euler's, 91
    Newton's, 51
    of action and reaction, 4, 51
    of cosines, 132
    of force and acceleration, 4, 51
    of gravitation, 5
    of inertia, 4, 51
    of motion, 4, 51, 91
    of sines, 132
    parallelogram, 133
    Pascal's, 32
length
    dimension of, 140
    natural, 6
lever arm, 8
line density, 23
line element, 23
line load, 26
line of action, 7
linear damper, 75
linear spring, 6
link, 50
liquid, 30
load
    distributed, 25
    line, 26
    surface, 25
local gravity constant, 5, 16
localization, 139

magnification factor, 78
magnitude
    of quantity, 140
    of vector, 133, 135
mass
    dimension of, 140
    of body, 21

of flat body, 94
mass element, 21, 94
mass moment of inertia, 94
material point, 2
mechanical energy, 60
meter (m), 140
mirror symmetry, 28
moment arm, 8
moment equation, 91–93, 118
    for planar motion, 98
moment equilibrium, 14
moment of force, 8
moment of inertia, 119
    def. of, 95
    table of, 144
    transfer of, 95
    transfer theorem for, 120
moment reference point, 8
moment sum
    def. of, 11
    transfer theorem of, 11
moment vector, 8
moment–angular momentum rela-
    tion, 65
momentum
    conservation of, 69
    of body, 90
    of particle, 63
    of rigid body, 92
    of system of particles, 66
motion, 51
    circular, 47
    curvilinear, 43
    harmonic, 74
    law of, 51, 91
    planar, 42, 82
    rectilinear, 42, 51
    rotational, 83
    three-dimensional, 111
    translational, 82
    uniform, 3, 51
multi-body system, 19, 38

natural angular frequency, 74
natural basis vector, 47
natural length, 6
natural period, 74
negation of vector, 133
newton (N), 4
Newton's law of gravitation, 5

Newton's laws of motion, 4, 51
normal force, 28, 36
normal unit vector, 48
numerical value, 142

oblique impact, 71
orientation, 83
orthogonal coordinate system, 134
orthonormal basis, 134
oscillator
  damped, 75
  free, 74
  harmonic, 73
  undamped, 73, 74
osculating circle, 48
overdamped, 76

parallel axis theorem, 121
parallelogram law, 133
particle, 2
particle dynamics, 41
particular solution, 77
pascal (Pa), 30
Pascal's law, 32
Pascal's principle, 32
period, 74
phase angle, 75, 133
physical dimension, 140
place of contact, 37
planar kinematics
  of particles, 42
  of rigid bodies, 82
planar kinetics
  of rigid bodies, 90
planar motion
  of particles, 42
  of rigid body, 82
planar section, 27, 30
Planck constant, 141
plane
  reference, 12, 66, 82, 96
  tangent, 36
plastic impact, 71
point
  body-fixed, 86
  material, 2
  moment reference, 8
  of action, 58
  of application, 7
  of reduction, 11

space-fixed, 58
point contact, 16, 27
polar angle, 83
polar basis vector, 45
polar coordinate, 45
polar coordinate system, 45
position vector, 3
potential energy
  of particle, 60
  of rigid body, 104
pound-force (lb_f), 4, 142
pound-mass (lb_m), 142
power
  of couple, 99
  of force, 57, 99
power sum, 100
precession, 130
prefix, 141
pressure, 30
principle
  Archimedes', 35
  Pascal's, 32
product
  cross, 135
  dot, 134
  scalar, 134
product of inertia, 119
  transfer theorem for, 120
product rule, 136
projection area, 31
psi, 31
pulley, 18, 50

quantity, 140

radian (rad), 132
radius of curvature, 48
radius vector, 3
reciprocal action, 3
rectangular basis vector, 43
rectangular coordinate, 43
rectangular coordinate system, 43
rectilinear motion, 42
reduced force-couple system, 11
reference plane
  of force-couple system, 12
  of planar motion, 66, 82, 96
relative angular velocity, 115, 116
relative-acceleration equation, 86, 115

relative-velocity equation, 85, 115
resonance, 79
resonance angular frequency, 79
rest, 14
restoring force, 73
resultant, 25
right triangle, 132
right-hand coordinate system, 134
right-hand rule, 8, 84, 134, 136
rigid body, 2
rigid body dynamics, 81
rolling without slipping, 88
rotating coordinate system, 113
rotation, 83
  fixed-axis, 88, 124
rotational motion, 83

scalar product, 134
second (s), 140
second-order differential equation, 76
section, 27, 30
section couple, 27
section force, 27
separable differential equation, 138
shear force, 28
shell, 23
SI system, 140
sliding, 37
sliding initiation, 37
slug (slug), 142
smooth surface, 36
space-fixed point, 58
speed, 42
speed of light, 51, 141
spin, 116, 128
spring, 6, 60
spring constant, 6
spring force, 6
static equilibrium, 14
static friction, 37
statics, 1, 4
  of fluids, 30
Steiner's theorem, 95
string, 17, 50
strong damping, 76
substitution, 139
subtraction of vectors, 133
support, 17
surface
  frictionless, 36

smooth, 36
surface density, 23, 94
surface element, 23, 94
surface load, 25
symmetry plane, 94, 119
system
    force-couple, 10
    inertial, 4, 52
    multi-body, 19, 38
    zero, 12, 14
system of particles, 66

tangent plane, 36
tangent unit vector, 48
tensile force, 18
terrestrial coordinate system, 52
thickness, 94
thickness direction, 94
three-dimensional kinematics, 111
three-dimensional kinetics, 118
three-dimensional motion, 111
time, 3, 140
time derivative, 3, 136
transcendental function, 143
transfer theorem
    for moment of inertia, 120
    for product of inertia, 120
    of angular momentum, 92
    of moment sum, 11
transfer vector, 120
transient, 77
translation, 82
    rectilinear, 14
translational motion, 82
trigonometric identity, 132
two-force member, 18

unbalance, 124

undamped oscillator, 73
underdamped, 76
uniform motion, 3, 51
unit, 140
    foot (ft), 142
    hertz (Hz), 74
    horsepower (hp), 57
    joule (J), 57
    kilogram (kg), 140
    meter (m), 140
    newton (N), 4
    pascal (Pa), 30
    pound-force (lb_f), 4, 142
    pound-mass (lb_m), 142
    psi, 31
    radian (rad), 132
    second (s), 140
    slug (slug), 142
    watt (W), 57
unit vector, 7, 134, 135
    normal, 48
    tangent, 48
universal constant, 140

vector
    addition of, 133
    body-fixed, 114
    component of, 134
    direction, 7, 134
    force, 7
    geometric, 133
    moment, 8
    position, 3
    radius, 3
    subtraction of, 133
    transfer, 120
    unit, 7, 134, 135
    zero, 133

vector negation, 133
vector-valued function, 136, 138
velocity, 3
    angular, 45
    def. of, 43
    in arc coordinates, 48
    in natural basis, 48
    in polar coordinates, 46
    in rectangular coordinates, 44
    instantaneous, 42
    relative, 85, 115
vibration, 79
volume element, 21

watt (W), 57
weak damping, 76
wedge, 39
wheel, 88
work
    of couple, 100
    of force, 57, 99
    of force of gravity, 60, 105
    of spring force, 61
    on particle, 58
    on rigid body, 102
work integral
    between orientations, 100
    between positions, 58
work–energy method
    for particles, 57
    for rigid bodies, 99
work–energy theorem
    for particles, 59, 62
    for rigid bodies, 104, 105

zero system, 12, 14
zero vector, 133

www.ingramcontent.com/pod-product-compliance
Lightning Source LLC
Chambersburg PA
CBHW051218200326
41519CB00025B/7160